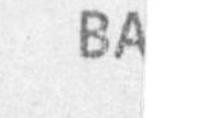

ABOUT THE AUTHORS

Nick Polson is Professor of Econometrics and Statistics at the Chicago Booth School of Business. He is a Bayesian statistician involved in research in machine intelligence, deep learning and computational methods. He regularly speaks to large audiences in the US, UK and the rest of Europe.

James Scott is Associate Professor of Statistics at the University of Texas. James is a statistician and data scientist who studies Bayesian inference and computational methods for big data. James lives in Austin, Texas with his wife, Abigail.

www.penguin.co.uk

AIQ

HOW ARTIFICIAL INTELLIGENCE WORKS AND HOW WE CAN HARNESS ITS POWER FOR A BETTER WORLD

NICK POLSON AND JAMES SCOTT

BLACK SWAN

TRANSWORLD PUBLISHERS
61–63 Uxbridge Road, London W5 5SA
www.penguin.co.uk

Transworld is part of the Penguin Random House group of companies
whose addresses can be found at global.penguinrandomhouse.com

Penguin
Random House
UK

First published in Great Britain in 2018 by Bantam Press
an imprint of Transworld Publishers
Black Swan edition published 2019

A CIP catalogue record for this book
is available from the British Library.

ISBN
9781784163372

Typeset in 10.37/13.8pt Minion Pro.
Printed and bound in Great Britain by Clays Ltd, Elcograf S.p.A.

fu

To Diana and to Anne.

—NP

To my grandparents, to Margaret Aiken, and
to the golden guinea.

—JS

CONTENTS

AIQ

INTRODUCTION

WE TEACH DATA science to hundreds of students per year, and they're all fascinated by artificial intelligence. And they ask *great* questions. How does a car learn to drive itself? How does Alexa understand what I'm saying? How does Spotify pick such good playlists for me? How does Facebook recognize my friends in the photos I upload? These students realize that AI isn't some sci-fi droid from the future; it's right here, right now, and it's changing the world one smartphone at a time. They all want to understand it, and they all want to be a part of it.

And our students aren't the only ones enthusiastic about AI. They're joined in their exaltation by the world's largest companies—from Amazon, Facebook, and Google in America to Baidu, Tencent, and Alibaba in China. As you may have heard, these big tech firms are waging an expensive global arms race for AI talent, which they judge to be essential to their future. For years we've seen them court freshly minted PhDs with offers of $300,000+ salaries and much better coffee than we have in academia. Now we're seeing many more companies jump into the AI recruiting fray—firms sitting on piles of data in, say, insurance or the oil business, who are coming along with whopping salary offers and fancy espresso machines of their own.

Yet while this arms race is real, we think there's a much more powerful trend at work in AI today—a trend of diffusion and dissemination, rather than concentration. Yes, every big tech company is trying to hoard math and coding talent. But at the same time, the underlying technologies and ideas behind AI are spreading with extraordinary speed: to smaller companies, to other parts of the economy, to hobbyists and coders and scientists and researchers everywhere in the world. That democratizing trend, more than anything else, is what has our students today so excited, as they contemplate a vast range of problems practically begging for good AI solutions.

Who would have thought, for example, that a bunch of undergraduates would get so excited about the mathematics of cucumbers? Well, they did when they heard about Makoto Koike, a car engineer from Japan whose parents own a cucumber farm. Cucumbers in Japan come in a dizzying variety of sizes, shapes, colors, and degrees of prickliness—and based on these visual features, they must be separated into nine different classes that command different market prices. Koike's mother used to spend *eight hours per day* sorting cucumbers by hand. But then Koike realized that he could use a piece of open-source AI software from Google, called TensorFlow, to accomplish the same task, by coding up a "deep-learning" algorithm that could classify a cucumber based on a photograph. Koike had never used AI or TensorFlow before, but with all the free resources out there, he didn't find it hard to teach himself how. When a video of his AI-powered sorting machine hit YouTube, Koike became an international deep-learning/cucumber celebrity. It wasn't merely that he had given people a feel-good story, saving his mother from hours of drudgery. He'd also sent an inspiring message to students and coders across the world: that if AI can solve problems in cucumber farming, it can solve problems just about anywhere.

That message is now spreading quickly. Doctors are now using AI to diagnose and treat cancer. Electrical companies use AI to improve power-generating efficiency. Investors use it to manage financial risk. Oil companies use it to improve safety on deep-sea rigs. Law enforcement agencies use it to hunt terrorists. Scientists use it to make new discoveries in astronomy and physics and neuroscience. Companies, researchers, and hobbyists everywhere are using AI in thousands of different ways, whether to sniff for gas leaks, mine iron, predict disease outbreaks, save honeybees from extinction, or quantify gender bias in Hollywood films. And this is just the beginning.

We see the *real* story of AI as the story of this diffusion: from a handful of core math concepts stretching back decades, or even centuries, to the supercomputers and talking/thinking/cucumber-sorting machines of today, to the new and ubiquitous digital wonders of tomorrow. Our goal in this book is to tell you that story. It is partly a story of technology, but it is mainly a story of ideas, and of the people behind those ideas—people from a much earlier age, people who were just keeping their heads down and solving their own problems involving math and data, and who had *no clue* about the role their solutions would come to play in inventing the modern world. By the end of that story, you'll understand what AI is, where it came from, how it works, and why it matters in your life.

What Does "AI" Really Mean?

When you hear "AI," don't think of a droid. Think of an *algorithm*.

An algorithm is a set of step-by-step instructions so explicit that even something as literal-minded as a computer can follow them. (You may have heard the joke about the robot who got stuck in the shower forever because of the algorithm on the shampoo bottle: "Lather. Rinse. Repeat.") On its own, an algorithm is no smarter than a power drill; it just does one thing very well, like sorting a list of numbers or searching the web for pictures of cute animals. But if you chain lots of algorithms together in a clever way, you can produce AI: a domain-specific illusion of intelligent behavior. For example, take a digital assistant like Google Home, to which you might pose a question like "Where can I find the best breakfast tacos in Austin?" This query sets off a chain reaction of algorithms:

One algorithm converts the raw sound wave into a digital signal.

Another algorithm translates that signal into a string of English phonemes, or perceptually distinct sounds: "brek-fust-tah-koze."

The next algorithm segments those phonemes into words: "breakfast tacos."

Those words are sent to a search engine—itself a huge pipeline of algorithms that processes the query and sends back an answer.

Another algorithm formats the response into a coherent English sentence.

A final algorithm verbalizes that sentence in a non-robotic-sounding way: "The best breakfast tacos in Austin are at Julio's on Duval Street. Would you like directions?"

And that's AI. Pretty much every AI system—whether it's a self-driving car, an automatic cucumber sorter, or a piece of software that monitors your credit card account for fraud—follows this same "pipeline-of-algorithms" template. The pipeline takes in data from some specific domain, performs a chain of calculations, and outputs a prediction or a decision.

There are two distinguishing features of the algorithms used in AI. First, these algorithms typically deal with probabilities rather than certainties. An algorithm in AI, for example, won't say outright that some credit card transaction is fraudulent. Instead, it will say that the probability of fraud is 92%—or whatever it thinks, given the data. Second, there's the question of how these algorithms "know" what instructions to follow. In traditional algorithms, like the kind that run websites or word processors, those instructions are fixed ahead of time by a programmer. In AI, however, those instructions are learned by the algorithm itself, directly from "training data." Nobody tells an AI algorithm how to classify credit card transactions as fraudulent or not. Instead, the algorithm sees lots of examples from each category (fraudulent, not fraudulent), and it finds the patterns that distinguish one from the other. In AI, the role of the programmer isn't to tell the algorithm what to do. It's to tell the algorithm how to *train itself* what to do, using data and the rules of probability.

How Did We Get Here?

Modern AI systems, like a self-driving car or a home digital assistant, are pretty new on the scene. But you might be surprised to learn that most of the big ideas in AI are actually *old*—in many cases, centuries old—and that our ancestors have been using them to solve problems for generations. For example, take self-driving cars. Google debuted its first such car in 2009. But you'll learn in chapter 3 that one of the main ideas behind how these cars work was discovered by a Presbyterian minister in the 1750s—and that this idea was used by a team of mathematicians over 50 years ago to solve one of the Cold War's biggest blockbuster mysteries.

Or take image classification, like the software that automatically tags your friends in Facebook photos. Algorithms for image processing have gotten radically better over the last five years. But in chapter 2, you'll learn that the key ideas here date to 1805—and that these ideas were used a century ago, by a little-known astronomer named Henrietta Leavitt, to help answer one of the deepest scientific questions that humans have ever posed: How big is the universe?

Or even take speech recognition, one of the great AI triumphs of recent years. Digital assistants like Alexa and Google Home are remarkably fluent with language, and they'll only get better. But the first person to get a computer to understand English was a rear admiral in the U.S. Navy, and she did so almost 70 years ago. (See chapter 4.)

Those are just three illustrations of a striking fact: no matter where you look in AI, you'll find an idea that people have been kicking around for a long time. So in many ways, the big historical puzzle isn't why AI is happening now, but why it didn't happen long ago. To explain this puzzle, we must look to three enabling technological forces that have brought these venerable ideas into a new age.

The first AI enabler is the decades-long exponential growth in the speed of computers, usually known as Moore's law. It's hard to convey intuitively just how fast computers have gotten. The cliché used to be that the Apollo astronauts landed on the moon with less computing power than a pocket calculator. But this no longer resonates, because . . . what's a pocket calculator? So we'll try a car analogy instead. In 1951, one of the fastest computers was the UNIVAC, which performed 2,000 calculations per second, while one of the fastest cars was the Alfa Romeo 6C, which traveled 110 miles per hour. Both cars and computers have improved since 1951—but if cars had improved at the same rate as computers, then a modern Alfa Romeo would travel at 8 million times the speed of light.

The second AI enabler is the *new* Moore's law: the explosive growth in the amount of data available, as all of humanity's information has become digitized. The Library of Congress consumes 10 terabytes of storage, but in 2013 alone, the big four tech firms—Google, Apple, Facebook, and Amazon—collected about 120,000 times as much data as this. And that's a lifetime ago in internet years. The pace of data accumulation is accelerating faster than an Apollo rocket; in 2017, more than 300 hours of video were uploaded to YouTube every minute, and more than 100 million images were posted to Instagram every day. More data means smarter algorithms.

The third AI enabler is cloud computing. This trend is nearly invisible to consumers, but it's had an enormous democratizing effect on AI. To illustrate this, we'll draw an analogy here between data and oil. Imagine if all companies of the early twentieth century had owned some oil, but they had to build the infrastructure to extract, transport, and refine that oil on their own. Any company with a new idea for making good use of its oil would have faced enormous fixed costs just to get started; as a result, most of the oil would have sat in the ground. Well, the same logic holds for data, the oil of the twenty-first century. Most hobbyists or small companies would face prohibitive costs if they had to buy all the gear and expertise needed to build an AI system from their data. But the cloud-computing resources provided by outfits like Microsoft Azure, IBM, and Amazon Web Services have turned that fixed cost into a variable cost, radically changing the economic calculus for large-scale data storage and analysis. Today, anyone who wants to make use of their "oil" can now do so cheaply, by renting someone else's infrastructure.

When you put those four trends together—faster chips, massive data sets, cloud computing, and above all *good ideas*—you get a supernova-like explosion in both the demand and capacity for using AI to solve real problems.

AI Anxieties

We've told you how excited our students are about AI, and how the world's largest firms are rushing to embrace it. But we'd be lying if we said that *everyone* was so bullish about these new technologies. In fact, many people are anxious, whether about jobs, data privacy, wealth concentration, or Russians with fake-news Twitter-bots. Some people—most famously Elon Musk, the tech entrepreneur behind Tesla and SpaceX—paint an even scarier picture: one where robots become self-aware, decide they don't like being ruled by people, and start ruling *us* with a silicon fist.

Let's talk about Musk's worry first; his views have gotten a lot of attention, presumably because people take notice when a member of the billionaire disrupter class talks about artificial intelligence. Musk has claimed that in developing AI technology, humanity is "summoning a demon," and that smart machines are "our biggest existential threat" as a species.

After you've read our book, you'll be able to decide for yourself whether you think these worries are credible. We want to warn you up front, however, that it's very easy to fall into a trap that cognitive scientists call the "availability heuristic": the mental shortcut in which people evaluate the plausibility of a claim by relying on whatever immediate examples happen to pop into their minds. In the case of AI, those examples are mostly from science fiction, and they're mostly evil—from the Terminator to the Borg to HAL 9000. We think that these sci-fi examples have a strong anchoring effect that makes many people view the "evil AI" narrative less skeptically than they should. After all, just because we can dream it and make a film about it doesn't mean we can build it. Nobody today has any idea how to create a robot with *general* intelligence, in the manner of a human or a Terminator. Maybe your remote descendants will figure it out; maybe they'll even program their creation to terrorize the remote descendants of Elon Musk. But that will be their choice and their problem, because no option on the table today even remotely foreordains such a possibility. Now, and for the foreseeable future, "smart" machines are smart only in their specific domains:

- Alexa can read you a recipe for spaghetti Bolognese, but she can't chop the onions, and she certainly can't turn on you with a kitchen knife.
- An autonomous car can drive you to the soccer field, but it can't even referee the match, much less decide on its own to tie you to the goalposts and kick the ball at your sensitive bits.

Moreover, consider the opportunity cost of worrying that we'll soon be conquered by self-aware robots. To focus on this possibility *now* is analogous to the de Havilland Aircraft Company, having flown the first commercial jetliner in 1952, worrying about the implications of warp-speed travel to distant galaxies. Maybe one day, but right now there are far more important things to worry about—like, to press the jetliner analogy a little further, setting smart policy for all those planes in the air *today.*

This issue of policy brings us to a whole other set of anxieties about AI, much more plausible and immediate. Will AI create a jobless world? Will machines make important decisions about your life, with zero accountability? Will the people who own the smartest robots end up owning the future?

These questions are deeply important, and we hear them discussed all the time—at tech conferences, in the pages of the world's major newspapers, and over lunch among our colleagues. We should let you know up front that you won't find the answers to these questions in our book, because we don't know them. Like our students, we are ultimately optimistic about the future of AI, and we hope that by the end of the book, you will share that optimism. But we're not labor economists, policy experts, or soothsayers. We're data scientists—and we're also academics, meaning that our instinct is to stay firmly in our lane, where we're confident of our expertise. We can teach you about AI, but we can't tell you for sure what the future will bring.

We *can* tell you, however, that we've encountered a common set of narratives that people use to frame this subject, and we find them all incomplete. These narratives emphasize the wealth and power of the big tech firms, but they overlook the incredible democratization and diffusion of AI that's already happening. They highlight the dangers of machines making important decisions using biased data, but they fail to acknowledge the biases or outright malice in *human* decision-making that we've been living with forever. Above all, they focus intensely on what machines may take away, but they lose sight of what we'll get in return: different and better jobs, new conveniences, freedom from drudgery, safer workplaces, better health care, fewer language barriers, new tools for learning and decision-making that will help us all be smarter, better people.

Take the issue of jobs. In America, jobless claims kept hitting new lows from 2010 through 2017, even as AI and automation gained steam as economic forces. The pace of robotic automation has been even more relentless in China, yet wages there have been soaring for years. That doesn't mean AI hasn't threatened *individual people's* jobs. It has, and it will continue to do so, just as the power loom threatened the jobs of weavers, and just as the car threatened the jobs of buggy whips. New technologies always change the mix of labor needed in the economy, putting downward pressure on wages in some areas and upward pressure in others. AI will be no different, and we strongly support job-training and social-welfare programs to provide meaningful help for those displaced by technology. A universal basic income might even be the answer here, as many Silicon Valley bosses seem to think; we don't claim to know. But arguments that AI will create a jobless future are, so far, completely unsupported by actual evidence.

Then there's the issue of market dominance. Amazon, Google, Facebook, and Apple are enormous companies with tremendous power. It is *critical* that we be vigilant in the face of that power, so that it isn't used to stifle competition or erode democratic norms. But don't forget that these companies are successful because they have built products and services that people love. And they'll only continue to be successful if they keep innovating, which isn't easy for large organizations. Besides, we've read a lot of predictions that the big tech firms of today will remain dominant forever, and we find that these predictions usually don't even explain the past, much less the future. Remember when Dell and Microsoft were dominant in computing? Or when Nokia and Motorola were dominant in cell phones—so dominant that it was hard to imagine otherwise? Remember when every lawyer had a BlackBerry, when every band was on Myspace, or when every server was from Sun Microsystems? Remember AOL, Blockbuster Video, Yahoo!, Kodak, or the Sony Walkman? Companies come and companies go, but time marches on, and the gadgets just keep getting cooler.

We take a practical outlook on the emergence of AI: it is here today, and more of it is coming tomorrow, whether any of us like it or not. These technologies will bring *immense* benefits, but they will also, inevitably, reflect our weak spots as a civilization. As a result, there will be dangers to watch out for, whether to privacy, to equality, to existing institutions, or to something nobody's even thought of yet. We must meet these dangers with smart policy—and if we hope to craft smart policy in a world of "hot takes" and 140 characters, it is essential that we reach a point as a society where we can discuss these issues in a balanced way, one that reflects both their importance and their complexity. Our book isn't going to present that discussion. But it *will* show you what you need to understand if you want to play an informed role in that discussion.

A Note on Math

Before we get started, we owe you one final heads-up: you'll encounter some math in this book. Even if you've never thought of yourself as a math person, please don't worry. The math of AI is *surprisingly* simple, and we promise you'll be up to it. We also promise that it will be worth the effort: if you understand a bit of the math behind AI, you'll find that it becomes a lot less mysterious.

We could have written a book about AI without any math whatsoever, on the theory we've heard our whole lives that you can choose math or friends but not both. Our editor initially *begged* us to take this approach, muttering something under his breath about "losing three thousand readers per math symbol," or maybe it was "five thousand readers per Greek letter." Whatever it was, we said no δαμυεδ way, because our experience has taught us that we can have much more faith in you than that. Between the two of us, we've been teaching data science and probability for 40 years, including to lots of MBAs and undergraduates who come to us preinstalled with this horrible virus that makes them actually *dislike math.* Yet we've seen how those same students' eyes light up when they learn how all the cool AI applications they've heard of, from Alexa to image recognition, really work—how, when you get right down to it, it's all just probability on big-data steroids. They come to understand that the equations were never all that hard in the first place. By the end, they even feel *empowered* by the math; they realize that in the right circumstances, thinking a bit more like a machine—that is, making decisions using data and the rules of probability—can even help you be a smarter person.

So come join us over the next seven chapters, as we introduce you to seven fascinating historical characters, each with an important lesson to teach you about why smart machines need smart people, and vice versa. You'll come away with a higher AIQ and a new appreciation for just how brilliant human beings can be when they put their minds and their technology together.

1

THE REFUGEE

On personalization: how a Hungarian émigré used conditional probability to protect airplanes from enemy fire in World War II, and how today's tech firms are using the same math to make personalized suggestions for films, music, news stories—even cancer drugs.

NETFLIX HAS COME so far, so fast, that it's hard to remember that it started out as a "machine learning by mail" company. As recently as 2010, the company's core business involved filling red envelopes with DVDs that would incur "no late fees, ever!" Each envelope would come back a few days after it had been sent out, along with the subscriber's rating of the film on a 1-to-5 scale. As that ratings data accumulated, Netflix's algorithms would look for patterns, and over time, subscribers would get better film recommendations. (This kind of AI is usually called a "recommender system"; we also like the term "suggestion engine.")

Netflix 1.0 was so focused on improving its recommender system that in 2007, to great fanfare among math geeks the world over, it announced a public machine-learning contest with a prize of $1 million. The company put some of its ratings data on a public server, and it challenged all comers to improve upon Netflix's own system, called Cinematch, by at least 10%—that is, by predicting how you'd rate a film with 10% better accuracy than Netflix could. The first team to meet the 10% threshold would win the cash.

Over the ensuing months, thousands of entries flooded in. Some came tantalizingly close to the magic 10% threshold, but nobody beat it. Then in 2009, after two years of refining their algorithm, a team calling themselves BellKor's Pragmatic Chaos finally submitted the million-dollar piece of code, beating Netflix's engine by 10.06%. And it's a good thing they didn't pause to watch an extra episode of *The Big Bang Theory* before hitting the submit button. BellKor reached the finish line of the two-year race just 19 minutes and 54 seconds ahead of a second team, The Ensemble, who submitted an algorithm also reaching 10.06% improvement—just not quite fast enough.

In retrospect, the Netflix Prize was a perfect symbol of the company's early reliance on a core machine-learning task: algorithmically predicting how a subscriber would rate a film. Then, in March of 2011, three little words changed the future of Netflix forever: *House of Cards.*

House of Cards was the first "Netflix Original Series," the company's first try at producing TV rather than merely distributing it. The production team behind *House of Cards* originally went to all the major networks with their idea, and every single one was interested. But they were all cautious—and they all wanted to see a pilot first. The show, after all, is a tale of lies, betrayal, and murder. You can almost imagine the big networks asking themselves, "How can we be sure that anyone will watch something so sinister?" Well, Netflix could. According to the show's producers, Netflix was the only network with the courage to say, "We believe in you. We've run our data, and it tells us that our audience would watch this series. We don't need you to do a pilot. How many episodes do you want to do?"[1]

We've run our data, and we don't need a pilot. Think of the economic implications of that statement for the television industry. In the year before *House of Cards* premiered, the major TV networks commissioned 113 pilots, at a total cost of nearly $400 million. Of those, only 35 went on the air, and only 13—one show in *nine*—made it to season two. Clearly the networks had almost no idea what would succeed.

So what did Netflix know in March of 2011 that the major networks didn't? What made its people so confident in their assessment that they were willing to move beyond *recommending* personalized TV and start *making* personalized TV?

The pat answer is that Netflix had data on its subscriber base. But while data was important, this explanation is far too simple. The networks had lots of data, too, in the form of Nielsen ratings and focus groups and countless thousands of surveys—and big budgets for gathering more data, if they believed in its importance.

The data scientists at Netflix, however, had two things that the networks did not, things that were just as important as the data itself: (1) the deep knowledge of probability required to ask the right questions of their data, and (2) the courage to rebuild their entire business around the answers they got. The result was an astonishing transformation for Netflix: from a machine-learning-powered distribution network to a new breed of production company in which data scientists and artists come together to make awesome television. As Ted Sarandos, Netflix's chief content officer, famously put it in an interview with *GQ*: "The goal is to become HBO faster than HBO can become us."[2]

Today, few organizations use AI for personalization better than Netflix, and the approach it pioneered now dominates the online economy. Your digital trail yields personalized suggestions for music on Spotify, videos on YouTube, products on Amazon, news stories from *The New York Times*, friends on Facebook, ads on Google, and jobs on LinkedIn. Doctors can even use the same approach to give you personalized suggestions for cancer therapy, based on your genes.

It used to be that the most important algorithm in your digital life was search, which for most of us meant Google. But the key algorithms of the future are about suggestions, not search. Search is narrow and circumscribed; you have to know what to search for, and you're limited by your own knowledge and experience. Suggestions, on the other hand, are rich and open ended; they draw on the accumulated knowledge and experience of billions of other people. Suggestion engines are like "doppelgänger software" that might someday come to know your preferences better than you do, at least consciously. How long will it be, for example, before you can tell Alexa, "I'm feeling adventurous; book me a weeklong holiday," and expect a brilliant result?

There's obviously a lot of sophisticated math behind these suggestion engines. But if you're math-phobic, there's also some very good news. It turns out that there's really only one key concept you need to understand, and it's this: to a learning machine, "personalization" means "conditional probability."

In math, a conditional probability is the chance that one thing happens, given that some other thing has already happened. A great example is a weather forecast. If you were to look outside this morning and see gathering clouds, you might assume that rain is likely and bring an umbrella to work. In AI, we express this judgment as a conditional probability—for example, "the conditional probability of rain this afternoon, given clouds this morning, is 60%." Data scientists write this a bit more compactly: $P(\text{rain this afternoon} \mid \text{clouds this morning}) = 60\%$. P means "probability," and that vertical bar means "given" or "conditional upon." The thing on the left of the bar is the event we're interested in. The thing on the right of the bar is our knowledge, also called the "conditioning event": what we believe or assume to be true.

Conditional probability is how AI systems express judgments in a way that reflects their partial knowledge:

> You just gave *Sherlock* a high rating. What's the conditional probability that you will like *The Imitation Game* or *Tinker Tailor Soldier Spy*?
>
> Yesterday you listened to Pharrell Williams on Spotify. What's the conditional probability that you'll want to listen to Bruno Mars today?
>
> You just bought organic dog food. What's the conditional probability that you will also buy a GPS-enabled dog collar?
>
> You follow Cristiano Ronaldo (@cristiano) on Instagram. What's the conditional probability that you will respond to a suggestion to follow Lionel Messi (@leomessi) or Gareth Bale (@garethbale11)?

Personalization runs on conditional probabilities, all of which must be estimated from massive data sets in which *you* are the conditioning event. In this chapter, you'll learn a bit of the magic behind how this works.

Abraham Wald, World War II Hero

The core idea behind personalization is a lot older than Netflix, older even than television itself. In fact, if you want to understand the last decade's revolution in the way that people engage with popular culture, then the best place to start isn't in Silicon Valley, or in the living room of a cord-cutting millennial in Brooklyn or Shoreditch. Rather, it's in 1944, in the skies over occupied Europe, where one man's mastery of conditional probability saved the lives of an untold number of Allied bomber crews in the largest aerial campaign in history: the bombardment of the Third Reich.

During World War II, the size of the air war over Europe was truly staggering. Every morning, vast squadrons of British Lancasters and American B-17s took off from bases in England and made their way to their targets across the Channel. By 1944, the combined Allied air forces were dropping over 35 million pounds of bombs *per week*. But as the air campaign escalated, so too did the losses. On a single mission in August of 1943, the Allies dispatched 376 bombers from 16 different air bases, in a joint bombing raid on factories in Schweinfurt and Regensburg in Germany. Sixty planes never came back—a daily loss of 16%. The 381st Bomb Group, flying out of RAF Ridgewell, lost 9 of its 20 bombers that day.[3]

World War II airmen were painfully aware that each mission was a roll of the dice. But in the face of these bleak odds, the bomber crews had at least three defenses.

1. Their own tail and turret gunners, to ward off attackers.
2. Their fighter escorts: the Spitfires and P-51 Mustangs sent along to defend the bombers from the Luftwaffe.
3. A Hungarian-American statistician named Abraham Wald.

Abraham Wald never shot down a Messerschmitt or even saw the inside of a combat aircraft. Nonetheless, he made an outsized contribution to the Allied war effort using an equally potent weapon: conditional probability. Specifically, Wald built a recommender system that could make personalized survivability suggestions for different kinds of planes. At its heart, it was just like a modern AI-based recommender system for TV shows. And when you understand how he built it, you'll also understand a lot more about Netflix, Hulu, Spotify, Instagram, Amazon, YouTube, and just about every tech company that's ever made you an automatic suggestion worth following.

Wald's Early Years

Abraham Wald was born in 1902 to a large Orthodox Jewish family in Kolozsvár, Hungary, which became part of Romania and changed its name to Cluj after World War I. His father, who worked at a bakery in town, created a home atmosphere of learning and intellectual curiosity for his six children. The young Wald and his siblings grew up playing the violin, solving math puzzles, and listening to stories at the feet of their grandfather, a famous and beloved rabbi. Wald attended the local university, graduating in 1926. He then went on to the University of Vienna, where he studied mathematics under a distinguished scholar named Karl Menger.[4]

By 1931, when he finished his PhD, Wald had emerged as a rare talent. Menger called his pupil's dissertation a "masterpiece of pure mathematics," describing it as "deep, beautiful, and of fundamental importance." But no university in Austria would hire a Jew, no matter how talented, and no matter how strongly his famous advisor recommended him. So Wald looked for other options. In fact, he told Menger that he was happy to take any job that would let him make ends meet; all he wanted to do was keep proving theorems and attending math seminars.

At first, Wald worked as the private math tutor for a wealthy Austrian banker named Karl Schlesinger, to whom Wald remained forever grateful. Then in 1933 he was hired as a researcher at the Austrian Institute for Business Cycle Research, where yet another famous scholar found himself impressed by Wald: economist Oskar Morgenstern, the coinventor of game theory. Wald worked side by side with Morgenstern for five years, analyzing seasonal variation in economic data. It was there at the institute that Wald first encountered statistics, a subject that would soon come to define his professional life.

But dark clouds were gathering over Austria. As Wald's advisor Menger put it, "Viennese culture resembled a bed of delicate flowers to which its owner refused soil and light, while a fiendish neighbor was waiting for a chance to ruin the entire garden." The spring of 1938 brought disaster: Anschluss. On March 11, Austria's elected leader, Kurt Schuschnigg, was deposed by Hitler and replaced by a Nazi stooge. Within hours, 100,000 troops from the German Wehrmacht marched unopposed across the border. By March 15 they were parading through Vienna. In a bitter omen, Karl Schlesinger, Wald's benefactor from the lean years of 1931–32, took his own life that very day.

Luckily for Wald, his work on economic statistics had earned attention abroad. The previous summer, in 1937, he'd been invited to America by an economics research institute in Colorado Springs. Although pleased by the recognition, Wald had initially been hesitant to leave Vienna. But Anschluss changed his mind, as he witnessed the Jews of Austria falling victim to a terrible orgy of murder and theft and betrayal. Their shops were plundered, their homes vandalized, their roles in public life stripped by the Nuremberg Laws—including Wald's role, at the Institute for Business Cycle Research. Wald was sad to say goodbye to Vienna, his second home, but he could see the winds of madness blowing stronger every day.

So in the summer of 1938, at great peril, he snuck across the border into Romania and traveled onward to America, dodging guards on the lookout for Jews fleeing the country. The decision to leave probably saved his life. Remaining in Europe were Wald's parents, his grandparents, and his five brothers and sisters—and all but one, his brother Hermann, were murdered in the Holocaust. By then Wald was living in America. He was safe and hard at work, married and with two children, and he took solace in the joys of his new life. Yet he would remain so stricken by grief over the fate of his family that he never again played the violin.

Wald in America

Abraham Wald would, however, do more than his fair share to make sure that Hitler faced the music.

The 35-year-old Wald arrived in America in the summer of 1938. Although he missed Vienna, he immediately liked his new home. Colorado Springs echoed the Carpathian foothills of his youth, and his new colleagues received him with warmth and affection. He didn't stay in Colorado for long, though. Oskar Morgenstern, who had fled to America himself and was now in Princeton, was telling his math friends all up and down the East Coast about his old colleague Wald, whom he described as a "gentle man" with "exceptional gifts and great mathematical power." Wald's reputation kept growing, and it soon caught the attention of an eminent statistics professor in New York named Harold Hotelling. In the fall of 1938, Wald accepted an offer to join Hotelling's group at Columbia University. He began as a research associate, but he flourished so rapidly as both a teacher and scholar that he was soon offered a permanent position on the faculty.

By late 1941, Wald had been in New York for three years, and the stakes of what was happening across the water were obvious to all but the willfully blind. For two of those years Britain had been fighting the Nazis alone, fighting, as Churchill put it, "to rescue not only Europe but mankind." Yet for those two long years, America had stood aside. It took the bombing of Pearl Harbor to rouse the American people from their torpor, but roused they were at last. Young men surged forward to enlist. Women joined factories and nursing units. And scientists rushed to their labs and chalkboards, especially the many émigrés who'd fled the Nazis in terror: Albert Einstein, John von Neumann, Edward Teller, Stanislaw Ulam, and hundreds of other brilliant refugees who gave American science a decisive boost during the war.

Abraham Wald, too, was eager to answer the call. He was soon given the chance, when his colleague W. Allen Wallis invited him to join Columbia's Statistical Research Group. The SRG had been started in 1942 by four statisticians who met periodically in a dingy room in Rockefeller Center, in midtown Manhattan, to provide statistical consulting to the military. As academics, they were initially unaccustomed to giving advice under pressure. Sometimes this led to episodes revealing comically poor perspective on the demands of war. In the SRG's early days, one mathematician complained resentfully about being forced by a secretary to save paper by writing his equations on both sides of the page.

But their gomer days didn't last long. By 1944, the Statistical Research Group had matured into a team of 16 statisticians and 30 young women from Hunter and Vassar Colleges who handled the computing work. The team became an indispensable source of technical advice to the military's Office of Scientific Research and Development, and their guidance was sought at the highest levels of command—and they got results. The statisticians at Columbia developed nothing so fearsome or famous as the teams gathered in Los Alamos or Bletchley Park at the same time. But their remit was broader, and their effect on the war was profound. They studied rocket propellants, torpedoes, proximity fuses, the geometry of aerial combat, the vulnerability of merchant vessels—anything involving math that would advance the war effort. As Wallis, the group's director, later reminisced:

> During the Battle of the Bulge in December 1944, several high-ranking Army officers flew to Washington from the battle, spent a day discussing the best settings on proximity fuses for air bursts of artillery shells against ground troops, and flew back to the battle.... This kind of responsibility, although rarely spoken of, was always in the atmosphere and exerted a powerful, pervasive, and unremitting pressure.[5]

Fortunately, it was a team of some of the best mathematical minds in the country, many of whom would go on to lead their chosen fields. Two became university presidents. Four served as president of the American Statistical Association. Mina Rees became the first female president of the American Association for the Advancement of Science. Milton Friedman and George Stigler received the Nobel Prize in Economics.

And on this team of all-stars, Abraham Wald was like LeBron James: the man who did everything. Only the hardest problems ever found their way to his desk, for even his fellow geniuses recognized that, in the words of the group's director, "Wald's time was too valuable to be wasted."

Wald and the Missing Airplanes

Wald's most famous contribution to the group's work was a paper that invented a branch of data analysis known as sequential sampling. His mathematical insights showed factories how they could produce fewer defective tanks and planes, just by implementing smarter inspection protocols. When this paper was declassified by the military, it made Wald an academic celebrity, and it changed the course of twentieth-century statistics, as researchers everywhere rushed to apply Wald's mathematical insights in new areas—especially in clinical trials, where those insights are still used today.

But our story here, about the exponential growth of Netflix-style personalization, relates to a different and almost universally misunderstood contribution of Abraham Wald's: his method for devising personalized survivability recommendations for aircraft.

Every day, the Allied air forces sent massive squadrons of airplanes to attack Nazi targets, and many planes returned having taken damage from enemy fire. At some point, someone in the navy had the clever idea of analyzing the distribution of hits on these returning planes. The thinking was simple: if you could find patterns in where the planes were taking fire, then you could recommend where to reinforce them with extra armor. These recommendations, moreover, could be personalized to each plane, since the threats to a nimble P-51 fighter were very different from the threats to a lumbering B-17 bomber.

The naïve strategy would be to put more armor wherever you saw lots of bullet holes on the returning planes. But this would be a bad idea, because the navy didn't have any data on the planes that got shot down. To see why this is so important, consider an extreme example. Suppose that a bomber could be shot down by a single hit to the engine, but that it was invulnerable to hits on the fuselage. If that were true, then the navy's data analysts would see hundreds of bombers coming back with harmless bullet holes on the fuselage—but not a single one coming back with holes around the engine, since every such plane would have crashed. Under this scenario, if you simply added armor where you saw the bullet holes—on the fuselage—then you'd actually be *handicapping* the bombers, adding weight that "protected" them from a nonexistent danger.

This example illustrates an extreme case of survivorship bias. Although the real world is much less extreme—bullets to the engine are not 100% lethal, nor are bullets to the fuselage 100% harmless—the statistical point remains: the pattern of damage on the returning planes had to be analyzed carefully.

At this juncture, we must pause to make two important side points. First, the internet bloody loves this story. Second, just about everyone who's ever told it—with the notable exception of an obscure, highly technical paper published in the *Journal of the American Statistical Association* in 1984—gets it wrong.[6]

Try Googling "Abraham Wald" and "World War II" yourself and see what you find: one blog post after another about how a mathematical crusader named Wald prevented those navy blockheads from making a terrible blunder and slapping a bunch of unnecessary armor on the fuselages of airplanes. We've read dozens of these things, and we have spared you the same dreary task by creating the following composite sketch.

> During World War II, the navy found a striking pattern of damage to planes returning from bombing runs in Germany, in which most of the bullet holes were on the fuselage. The navy guys reached the obvious conclusion: put more armor on the fuselage. Nonetheless, they gave their data to Abraham Wald, just to double-check. Wald's little gray cells went to work. And then a thunderbolt. "Wait!" Wald exclaims. "That's wrong. We don't see any damage to the engines because the planes that are hit in the engine never return. You need to add armor to the engine, not the fuselage." Wald had pointed out the crucial flaw in the navy's thinking: survivorship bias. His final, life-saving advice ran exactly counter to that of the other so-called experts: *put the armor where you don't see the bullet holes.*

We can see why this version of the story is so irresistible: the path of counterintuition eventually turns a full 360 degrees. Imagine asking any person off the street, "Where should we put extra armor on airplanes to help them survive enemy fire?" While we haven't done this survey, we suspect that "the engine" would be a popular response. But a naïve interpretation of the data initially seems to suggest otherwise: if the returning planes have taken damage on the fuselage, then by God, let's put the armor there instead. Only a genius like Wald, the story goes, can see to the heart of the matter, leading us back to our initial, intuitive conclusion.

Alas, as far as we can tell from the historical record, this account has little basis in fact. Worse still, this embellished version, in which the moral of the story is about survivorship bias, misses the truly important thing about Abraham Wald's contribution to the Allied war effort. Survivorship bias in the data was obviously the problem, and everybody knew it. Otherwise there would have been no reason to call the Statistical Research Group in the first place; the navy didn't need a bunch of math professors just to count bullet holes. Their question was more specific: how to estimate the conditional probability of an aircraft surviving an enemy hit in a particular spot, despite the fact that much of the relevant data was missing. The navy folks didn't know how to do this. They were really smart, but it is no insult to say that they weren't as smart as Abraham Wald.

Wald's real contribution was far subtler and more interesting than delivering some survivorship-bias boom-shakalaka to a cartoonish dolt of a navy commander. His masterstroke wasn't to identify the problem but to invent a solution: a "survivability recommender system," or a method that could provide military commanders with bespoke suggestions about how to improve survivability for *any* model of aircraft, using data on combat damage. Wald's algorithm was, in the words of the Statistical Research Group's director, an "ingenious piece of work by one of the greatest figures in the history of American statistics." Although Wald's algorithm wasn't published until the 1980s, it was used behind the scenes in World War II and for many years thereafter.[7] In the Vietnam War, the navy used Wald's algorithm on the A-4 Skyhawk; years later, the air force used it to improve the armor on the B-52 Stratofortress, the longest-serving aircraft in U.S. military history.

Missing Data: What You Don't Know Can Fool You

As you can now appreciate, Abraham Wald's problem of improving aircraft survivability was a whole lot like Netflix's problem of making personalized film suggestions. But there's a catch, and it's a big one.

> U.S. Navy, in 1943: "We want to estimate the conditional probability that a plane will crash, given that it takes enemy fire in a particular location, in light of the damage data from all other planes. This will allow us to personalize survivorship recommendations for each model of plane. But much of the data is missing: the planes that crash never return."
>
> Netflix, 70 years later: "We want to estimate the conditional probability that a subscriber will like a film, given his or her particular viewing history, in light of the ratings data from all other subscribers. This will allow us to personalize film recommendations for each viewer. But much of the data is missing: most subscribers haven't watched most films."

The catch is that both Abraham Wald and Netflix needed to estimate a conditional probability, but both faced the problem of missing data. And sometimes what's missing can be very informative.

Consider, for example, something that happened when one of your authors (Polson, a Brit) visited the other (Scott, a Texan) down in Austin for the first time. On a walk to a local coffee shop, we noticed a large white van parked on the street that read:

ARMADILLO
PET CARE

Imagine Polson's bemusement at the idea of a flourishing local business devoted to the needs of these very non-British creatures. What were armadillos like as pets? Did they learn their names? And why such a big van?

But then a delivery guy moved a trolley stacked high with packages from beside the van, and the quotidian truth was revealed:

ARMADILLO
CARPET CARE

Sometimes the missing part of the data changes the entire story.

It was just the same with Abraham Wald's data on aircraft survivability. Although his raw figures are lost to history, we can use his published navy report to hypothesize what he might have seen. Let's imagine following in Wald's footsteps as he examines the data on the Schweinfurt–Regensburg raid in August of 1943, where the Allies lost 60 of their 376 planes in a single day. The raw reports from the field would have looked something like this, where a question mark means "missing data":

Plane	Type of damage	Mission result
1) Hellcat Agnes	Fuselage	Returned home
2) The Bronx Bomber	?	Shot down
3) Pistol Pack'n Papa	Engine	Returned home
. . .	. . .	. . .
375) Homesick Angel	?	Shot down
376) Calamity Jane	None	Returned home

From these reports, Wald could have cross-tabulated the planes according to damage type and mission result.* This would have produced the following table:

Type of damage suffered	Returned (316 total)	Shot down (60 total)
Engine	29	?
Cockpit	36	?
Fuselage	105	?
None	146	0

Of the 316 planes making it back home, 105 have taken damage on the fuselage. This fact would have allowed Wald to estimate the condi-

* For you spreadsheet wizards, this is like making a pivot table.

tional probability that a plane has taken damage on the fuselage, given that it returns safely:

$$P(\text{damage on fuselage} \mid \text{returns safely}) = 105/316 \approx 32\%.$$

But that's the right answer to the wrong question. Instead, what we want to know is the exact inverse: the conditional probability that a plane returns safely, given that it has taken damage on the fuselage. This might be a very different number.

This brings us to an important rule about conditional probabilities: they are not symmetric. Just because Wald knew *P*(damage on fuselage | returns safely), he didn't necessarily know the inverse probability, *P*(returns safely | damage on fuselage). To illustrate why not, consider a simple example:

- All NBA players practice basketball, which means that *P*(practices basketball | plays in NBA) is nearly 100%.
- A vanishingly small fraction of those who practice basketball will make the NBA, which means *P*(plays in NBA | practices basketball) is nearly 0%.

So *P*(practices basketball | plays in NBA) does not equal *P*(plays in NBA | practices basketball). When thinking about probabilities, it's very important to be clear about which event is on the left side of the bar, and which event is on the right side of the bar.

Wald knew this. He knew that to calculate a probability like *P*(plane returns safely | damage on fuselage), he needed to estimate how many planes had taken damage to the fuselage and *never made it home*. His task was to put actual numbers in place of those question marks in the table above: that is, to fill in the missing data by reconstructing the statistical signature of the downed planes. Data scientists call this process "imputation." It's usually a lot better than "amputation," which means just chopping off the missing data.

Wald's attempt at imputation all came down to his modeling assumptions. He had to re-create the typical encounter of a B-17 with the enemy, using only the mute testimony of the bullet holes on the planes that had made it back, coupled with a hypothetical model of an aerial battle. To ensure that his modeling assumptions were as realistic as possible, Wald set to work like a forensic scientist. He analyzed the likely attack angle of enemy fighters. He chatted with engineers. He studied the properties of a shrapnel cloud from a flak gun. He even suggested that the army fire thousands of dummy bullets at a plane so that he could tabulate the hits.

When all was said and done, Wald had invented a method for reconstructing the full table. Based on his model of aerial battles, his estimates would have looked something like this:

Type of damage suffered	Returned (316 total)	Shot down (60 total)
Engine	29	31
Cockpit	36	21
Fuselage	105	8
None	146	0

From a filled-in data set like this one, it is now straightforward to estimate the conditional probabilities that Wald needed. For example, of 113 planes with hits to the fuselage, 105 of them returned home, and an estimated 8 didn't. Thus the conditional probability of returning safely, given damage to the fuselage, is

$$P(\text{plane returns safely} \mid \text{damage to fuselage}) = \frac{105}{105+8} \approx 93\%.$$

According to this estimate, a B-17 was very likely to survive a hit on the fuselage.

On the other hand, of the 60 planes that took damage to the engine, only 29 returned safely. Therefore

$$P(\text{returns safely} \mid \text{damage to engine}) = \frac{29}{29+31} \approx 48\%.$$

The bombers were much more likely to get shot down if they took damage to the engine.

These, finally, were the kind of numbers the navy could use. But more than just the numbers for a specific plane, it could also use Wald's approach to personalize the survivability recommendations for *any* plane. Conditional probability plus careful modeling of missing data proved to be a lifesaving combination.

Missing Bombers, Missing Ratings

Seventy years later, these same ideas would play a fundamental role in the way that Netflix reinvented itself as a company.

It all started from Netflix 1.0's recommender system, which we'll explain here in broad strokes. Imagine that you face the daunting task of designing this system yourself. As an input, the system must accept a subscriber's viewing history, and as an output, it must produce a prediction about whether that subscriber will like a particular show. You decide to start with an easy case inspired by Wald: assessing how probable it is that a subscriber will like the film *Saving Private Ryan*, given that he or she liked the HBO series *Band of Brothers*. This seems like a good bet: both are epic dramas about the Normandy invasion and its aftermath.

For this particular pair of shows, fine: recommend away. Keep in mind, however, that you want to be able to do this automatically. It would surely not be cost-effective to place a huge team of human annotators into the recommendation loop here, laboriously tagging all possible pairs of movies for similarities. But now recall that you have access to the entire Netflix database showing which customers have liked which films. Your goal is to leverage this vast data resource to automate the recommender system.

The key insight is to frame the problem in terms of conditional probability. Suppose that, for some pair of films A and B, the probability *P*(random subscriber likes film A | same random subscriber likes film B) is high—say, 80%. Now we learn from Linda's viewing history that she liked film B but hasn't yet seen film A. Wouldn't film A be a good recommendation? Based on her liking of B, there's an 80% chance she'll like A.

But how can we learn a number like *P*(subscriber likes *Saving Private Ryan* | subscriber likes *Band of Brothers*)? This is where your database comes in handy. To keep the numbers simple, let's say there are 100 people in your database, and every one of them has seen both films. Their viewing histories come in the form of a big "ratings matrix," where the rows correspond to subscribers and the columns to films:

Subscriber	Liked *Saving Private Ryan*?	Liked *Band of Brothers*?
1. Aaron	Yes	Yes
2. Alice	Yes	Yes
...	...	...
99. Wendy	No	No
100. Zack	Yes	No

Next, you cross-tabulate the data from the ratings matrix by counting how many subscribers had a specific combination of preferences for these two films:

	Liked *Band of Brothers*	Didn't like it
Liked *Saving Private Ryan*	56 subscribers	6 subscribers
Didn't like it	14 subscribers	24 subscribers

From this table, we can easily work out the conditional probability that your recommender system needs:

- 70 subscribers liked *Band of Brothers* (56 + 14).
- Of

these 70 subscribers, 56 of them liked *Saving Private Ryan*, and 14 didn't.

This allows you to calculate the conditional probability that someone who liked *Band of Brothers* will like *Saving Private Ryan*:

$$P(\text{likes } \textit{Saving Private Ryan} \mid \text{likes } \textit{Band of Brothers}) = \frac{56}{56+14} = 80\%.$$

The key thing that makes this approach work so well is that it's automatic. Computers aren't very good (yet) at automatically scanning films for thematic content. But they're brilliant at counting—that is, cross-tabulating vast databases of subscribers' movie-watching histories from a ratings matrix to estimate conditional probabilities.

The real problem that Netflix faces is much harder than this toy example, for at least three reasons. The first is scale. Netflix doesn't have 100 subscribers, it has 100 million, and it doesn't have ratings data on two shows, but on more than 10,000. As a result, the ratings matrix has more than a *trillion* possible entries.

The second issue is "missingness." Most subscribers haven't watched most films, so most of those trillion-plus entries in the ratings matrix are missing. Moreover, as in the case of the World War II bombers, that missingness pattern is informative. If you haven't watched *Fight Club*, maybe you just haven't gotten around to it—but then again, maybe films about nihilism just do nothing for you.

The final issue is combinatorial explosion. Or, if you'd rather stick with *Fight Club* and philosophy over mathematics: each Netflix subscriber is a beautiful and unique phenomenological snowflake. In a database with only two films, millions of users will share identical like/dislike experiences, since only four such experiences are possible: liked both, liked neither, or liked one but not the other. Not so in a database with 10,000 films. Consider your own film-watching history. No one else's history is exactly the same as yours, and no one else's ever will be, because there are too many ways to differ. Even in a database with only 300 films in it, there would be vastly more possible combinations of liking or disliking those films (2^{300}) than there are atoms in the universe (about 2^{272}). Long before you get to $2^{10,000}$, you might as well stop counting—the varieties of film-liking experience are, for all practical purposes, infinite.

This raises an important question. How can Netflix make a recommendation on the basis of your viewing history, using other people's viewing histories, when yours is unprecedented and theirs will never be repeated?

The solution to all three issues is careful modeling. Just as Wald solved his missing-data problem by building a model of a B-17's encounter with an enemy fighter, Netflix solved its problem by building a model of a subscriber's encounter with a film. And while Netflix's current model is proprietary, the million-dollar model built by team BellKor's

Pragmatic Chaos, winner of the Netflix Prize, is posted for free on the web.[8] Here's the gist of how it works. (Remember, Netflix predicts ratings on a 1-to-5 scale, from which a like/dislike prediction can be made using a simple cutoff, e.g., four stars.)

The fundamental equation here is

Predicted Rating = Overall Average + Film Offset + User Offset
+ User-Film Interaction.

The first three pieces of this equation are easy to explain.

- The overall average rating across all films is 3.7 stars.
- Every film has its own offset. *Schindler's List* and *Shakespeare in Love* have positive offsets because they're popular, while *Daddy Day Care* and *Judge Dredd* have negative offsets because they're not.
- Every user has an offset, because some users are more or less critical than average. Maybe Vladimir is a cynic and rates every film harshly (negative offset), while Donald thinks all films are terrific and rates them highly (positive offset).

These three terms provide a baseline rating for a given user/film pair. For example, imagine recommending *The Spy Who Loved Me* (film offset = 0.4) to Vlad the curmudgeon (user offset = −0.2). Vlad's baseline rating would be 3.7 + 0.4 − 0.2 = 3.9.

But that's just the baseline. It ignores the user-film interaction, which is where most of the data-science action happens. To estimate this interaction, the prizewinning team built something called a "latent feature" model. ("Latent feature" just means something not directly measured.) The idea here is that a person's ratings of similar films exhibit patterns because those ratings are all associated with latent features of that person. Each person's latent features can be estimated from prior ratings and used to make predictions about as-yet-unseen data. This same idea comes up everywhere, under many different names:

- Survey respondents give similar answers to questions about their job and education. Both are related to a latent feature, "socioeconomic status," that can also be used to predict a respondent's answer to a question about income. Social scientists call this "factor analysis."
- Senators vote in similar ways on taxes and health-care policy. Both are related to a latent feature, "ideology," that can also be used to predict a senator's vote on defense spending. Political scientists call this an "ideal point model."
- SAT test takers give similar patterns of answers to questions about geometry and algebra. Both are related to a latent feature, "math skill," that can also be used to predict a student's answer to a question about trigonometry. Test designers call this "item-response theory."
- Netflix subscribers rate *30 Rock* and *Arrested Development* in similar ways. Both are related to a latent feature—let's call it "affinity for witty oddball comedies"—that can also be used to predict a user's rating for *Parks and Recreation.* Data scientists call this "user-based collaborative filtering."

Of course, there's not just one latent feature to describe Netflix subscribers, but dozens or even hundreds. There's a "British murder mystery" feature, a "gritty character-driven crime drama" feature, a "cooking show" feature, a "hipster comedy films" feature, and so on. These features form the coordinate axes of a giant multidimensional space in which every user occupies a unique position, corresponding to the user's unique mix of preferences. Love *Poirot* but can't handle the violence of *Narcos*? Maybe you're a +2.5 on the British-murder-mystery axis and a –2.1 on the crime-drama axis. Adore *The Royal Tenenbaums* but find *The Great British Baking Show* a snooze? Maybe you're a 3.1 on hipster comedy and a –1.9 on cooking shows.

The coolest part about this whole process is that the latent features defining these axes aren't decided upon ahead of time. Instead, they are discovered organically by AI, using the patterns of correlation in tens of millions of user ratings. The data—not a critic or a human annotator—determines which shows go together.

The Hidden Features Tell the Story

We can now, at last, finish our story of personalization in AI. It is the story of how subscriber-level latent features, discovered from massive data sets using conditional probability, were the hidden force behind Netflix's strategic transformation from distributor to producer. It is also the story of how these latent features are the magic elixir of the digital economy—a special brew of data, algorithms, and human insight that represents the most perfect tool ever conceived for targeted marketing. The people who run Netflix realized this. They decided to use that tool to start making television shows themselves, and they never looked back.

Think about what makes Netflix different as a content producer. Unlike the major TV networks, Netflix doesn't care how old you are, what ethnicity you are, or where you live. It doesn't care about your job, your education, your income, or your gender. And it certainly doesn't care what the advertisers think, because there aren't any. The only thing Netflix cares about is what TV shows you like—something it understands in extraordinary detail, based on its estimates of your latent features.

Those features allow Netflix to segment its subscriber base according to hundreds of different criteria. Do you like dramas or comedies? Are you a sports fan? Do you like cooking shows? Do you like musicals? Do you like films with a more diverse cast? Do you watch every second of action films, or do you fast-forward when the violent parts come on? Do you watch cartoons? The patterns in your own viewing history, together with those learned from everyone else's history, give a mathematically precise answer to each of these questions, and to hundreds more. Your precise combination of latent features—your tiny little corner of a giant multidimensional Euclidean space—makes you a demographic of one.

And that's how Netflix invented its new business model of commissioning fantastic stories from world-class artists—some of them aimed at one miniaudience, and some at another. A great example is *The Crown,* an opulent, layered drama about the early life of Queen Elizabeth II. As of 2017, *The Crown* was the most expensive TV series ever made: $130 million for 10 episodes. Included in that budget were 7,000 period costumes, most famously a $35,000 royal wedding dress. It may sound as though Netflix is spending money like a drunken sailor on new programs. But remember those gory statistics from a single year of network television: $400 million commissioning 113 pilots, of which only 13 shows made it to a second season. When the standard industry practice is to blow hundreds of millions of dollars on shows destined for irrelevance, even a wedding dress that costs 300 annual Netflix subscriptions starts to look like a bargain. So rather than a drunken sailor, the better metaphor is a fortune-teller with a crystal ball—a data-driven, probabilistic crystal ball capable of telling the folks at Netflix exactly what kind of program their subscribers would pay $130 million for. Once they've figured that out, they trust the artists to do the rest.

The numbers are even starting to show that this approach works. Netflix doesn't release viewership statistics, but we do have at least one metric, and that's awards. In 2015, Netflix was in sixth place among TV networks for Emmy nominations. By 2017, it was in second place; its 91 nominations lagged behind only HBO's haul of 110, and HBO is justifiably worried about what will happen when the wildly popular *Game of Thrones* ends its run. It seems only a matter of time before streaming services like Netflix dominate the awards circuit.

Either way, the Netflix approach to personalization already dominates the digital economy. If the future of digital life is about suggestions rather than search, as we believe it is, then the future is also, inevitably, about conditional probability.

The Mixed Legacy of Suggestion Engines

Suggestion engines have been a major area of research in AI for a decade or more, both in academia and industry. Even though that legacy is still unfolding, it's worth reflecting on where we are now. The news is mixed.

The Dark Side of Targeted Marketing

First the bad news: these technologies haven't *only* been used to make suggestions about fun stuff, like TV shows and music. Suggestion engines also have a dark side, one that's been exploited in cynical, divisive ways. There's no better example than Russian agents' use of Facebook in the months leading up to America's 2016 presidential election.

Facebook is popular among advertisers for the same reason Netflix is popular among TV watchers: it's mastered the art of targeted marketing based on your digital trail. In ages past, when companies wanted to reach some demographic—college students, for example, or parents with school-age kids—those companies would buy ads in places where their target audience *might* be paying attention. Marketers made these ad-buying decisions using aggregate data on which people tended to watch this show or read that magazine. But they couldn't target specific individuals. As the marketer's old saw goes: half of all advertising dollars are wasted; we just don't know which half.

But if an ad used to be a blunt instrument, today it is a laser beam. Marketers can now design an online ad for any target audience they can imagine, defined at a level of demographic and psychographic detail that would boggle your mind. If you came to Facebook's sales team with a goal of targeting a vague group like "young professionals," for example, you would probably be laughed at behind your back. Tell us who you *really* want to target, you'd be told. Do you want lawyers or bankers? Democrats or Republicans? Sports fans or opera connoisseurs? Black or white, man or woman, North or South, steak or salad—and if salad, iceberg or kale? The list goes on and on. Once you've decided upon your audience, Facebook's algorithms can pick out *exactly* which users to target, and they can serve up an ad or a sponsored post to those users at exactly the moment when they're most likely to be receptive to its message. This makes marketers giddy—and it's why Facebook was, at the time of writing, worth over half a trillion dollars, more than the GDP of Sweden.

This kind of targeted marketing has been going on for a while, and judging by their behavior, most Facebook users have been willing to accept the "data for gossip" bargain implicit in their continued use of the platform. But for many people the alarm bells started to go off in the wake of the 2016 presidential election, when it became clear just how cleverly Russia had exploited Facebook's ad-targeting system to sow discord among American voters. Russian agents, for example, zeroed in on a group of users who'd gone on Facebook to express solidarity with police officers in the wake of protests by the Black Lives Matter movement. They targeted these users with an ad containing a picture of a flag-draped coffin at a policeman's funeral, along with the caption: "Another gruesome attack on police by a BLM movement activist. Our hearts are with those 11 heroes." They targeted a group of conservative Christian users with a different ad: a photo of Hillary Clinton shaking hands with a woman in a headscarf, together with a caption rendered in pseudo-Arabic script, "Support Hillary. American Muslims." The Russians made different ads for New Yorkers and for Texans, for LGBTQ advocates and for NRA supporters, for veterans and for civil rights activists—all of them targeted with ruthless algorithmic efficiency.[9]

We don't know anyone, of any party or any profession, who isn't appalled by the idea of a hostile foreign power weaponizing social media to influence an American election. And it is clear that the technology behind suggestion engines was at least one ingredient in this toxic cocktail of Russian money and identity politics. Once you move away from those points of near-universal agreement, however, the questions get a lot more complex. For example:

1. Did these activities change the outcome of the presidential election? We'll probably never know, since we cannot reverse engineer the decision-making processes of the 138.8 million people who cast a ballot, or the tens of millions more who stayed home.
2. Would it be different if a U.S.-based actor—say, the Koch brothers on the right, or the Blue Dog PAC on the left—did something similar? If you think that would still be objectionable, and that Facebook or someone else should put a stop to it, would you trust your political opponents to decide exactly where the limits should be?
3. Are these digital-age techniques qualitatively more effective at influencing people than the techniques that other propagandists, from Leni Riefenstahl to talk radio, have been using for years? This is a simple empirical question: If people are targeted with hyperspecific digital ads based on what the data says about them, and if that data is very accurate, how many change their minds or behave differently? If the answer is that the ads don't change any

minds but just make people more likely to vote the positions they already hold—again, assuming an American ad buyer—is that good or bad for democracy?

4. What, exactly, should be done now? Algorithms surely played a role in the Russia/Facebook debacle, but so did our preexisting political culture, and so did our laws on advertising, especially paid political advertising. What is the right mix of legal and policy responses to keep this kind of thing from happening again? Going further, should this be a wake-up call on targeted digital marketing *period,* not just in politics?

We don't know the answers to these questions, but we do believe that it's possible to have an informed conversation about them. So in the service of that conversation, here are two things to consider.

First, we can think of no clearer example than Russia's abuse of Facebook for why society cannot rely on machine intelligence without human supervision and still expect to see a brighter future. Suggestion engines are not going away, and there is no choice but to create a cultural and legal framework of oversight in which they can be used responsibly. We're optimistic that, given the chance, people *can* be smart enough to prevent the worst technological abuses without simply smashing all the machines.

Second, we can think of no better reason than this conversation why every citizen of the twenty-first century must understand some basic facts about artificial intelligence and data science. If education, as Thomas Jefferson said, is the cornerstone of democracy, then when it comes to digital technology, our democratic walls are falling down. Americans have debated the limits of commercial free speech almost since our birth as a country. But today we are *way* beyond a conversation about ads for sugary cereals on Saturday morning cartoons—far from the only example of a dubious marketing practice made utterly quaint by our new technology. There are *so many* unknowns that await us down the road. At a minimum, courts and legislatures should know more about their own blind spots, and stop dismissing details they don't understand as "gobbledygook." And citizens should participate in these discussions from a position of knowledge, rather than fear, of the basic technical details. Put simply, smart people who care about the world simply *must* know more about AI. That's one reason we wrote this book.

The Bright Side for Science

Now for some good news about suggestion engines. The mathematical and algorithmic insights produced by the last decade of work on personalization are just starting to bleed over into other areas of science and technology. As that happens, a lot of good things are in store for us.

Take the case of patient-centered social networks—like Crohnology, for people with gastrointesintal disorders; Tiatros, for soldiers with post-traumatic stress disorder; or PatientsLikeMe, for pretty much anything. These networks run on personalization algorithms, too, just like Facebook. Patients see them as an important resource for suggestions about treatments and lifestyle changes, while researchers see them as a valuable repository of real-world medical data that can be used to make those suggestions even better.

Or consider the expanding statistical toolkit of neuroscientists, who can now routinely monitor the activity of hundreds of neurons at once as they try to understand how the brain processes information. Hardware advances will soon enable them to monitor thousands of neurons or more. As their data sets grow larger and larger, neuroscientists are increasingly turning to Netflix-style latent-feature models to find clusters of neurons that tend to fire together in response to some stimulus—the neurophysiological equivalent of liking the same TV show. This work could lead to new discoveries and pioneering treatments for some of our most common maladies, from autism to Alzheimer's.

Perhaps the most exciting work of all is happening in cancer research, specifically something called "targeted therapy." While cancer may be labeled according to body part, fundamentally it is a disease of your genome. Tumor genomes, moreover, vary widely. Even patients with the same kind of cancer may have tumors with different genetic subtypes, and researchers have discovered that these subtypes often respond to drugs in very different ways. It is now common for doctors to test a sample of a patient's tumor for specific genes and proteins and to choose a cancer drug accordingly.

Over the years, cancer researchers have built large databases of genetic information on different tumor types, and they have joined forces with data scientists to mine those databases in search of patterns that can be exploited by targeted therapies. For example, about 60% of colorectal tumors have the wild-type (nonmutated) version of the KRAS gene. One particular cancer drug, cetuximab, is effective against these tumors but ineffective against the 40% that have a KRAS mutation.

That's a simple pattern, involving just one gene. Other patterns, on the other hand, are very complex. They involve dozens or hundreds of genes related to one of the many intricate molecular signaling pathways that go awry in cancer cells. To handle that complexity, researchers are increasingly turning to big-data latent-feature models, of the kind pioneered in Silicon Valley over the last decade to power large-scale recommender systems. These models are being used to analyze genomic data for an explanation of why some cancer patients respond to a drug and others don't. Just as Netflix uses the features in subscribers' viewing profiles to target them with TV shows, cancer researchers hope to use the features in patients' "genomic profiles" to target them with therapies—and maybe even design new ones for them, *House of Cards* style.

This idea is catching on. For example, in 2015, scientists at the National Cancer Institute announced that they had discovered two distinct subtypes of diffuse large B-cell lymphoma, based on latent genomic features. The scientists conjectured that the two subtypes, ABC and GCB, might respond differently to a particular drug, ibrutinib. So they enrolled 80 lymphoma patients in a clinical trial, took samples of their tumors to determine whether they were of subtype ABC or GCB, gave them all ibrutinib, and followed their progress over the ensuing months and years. The results were striking: ibrutinib was seven times more likely to work in patients with the ABC subtype.[10]

Given the long time horizon and billions of dollars required to develop and test a new cancer drug, this genomic-profiling strategy is far from mature. But as the ibrutinib trial shows, conditional probability is beginning to pay dividends in cancer research, and labs around the world are hard at work on the latest generation of targeted therapies.

Postscript

We hope this chapter has helped you to understand a bit more about the core idea behind companies like Netflix, Spotify, and Facebook: that, to a machine, "personalization" means "conditional probability." We also hope you've come to appreciate that these modern AI systems represent just one step along a winding historical path of human ingenuity—a path that will surely lead to new wonders, but one fraught with new challenges, too.

To close this chapter, we'll leave you with one last suggestion-engine story of our own. Over the summer of 2014, one of your authors (Scott) visited Ypres, a town in western Belgium whose strategic position loomed large in the early days of World War I. The German and Allied armies met outside Ypres in October of 1914. Both sides dug trenches, and a brutal years-long stalemate ensued:

> Men marched asleep. Many had lost their boots
> But limped on, blood-shod. All went lame; all blind;
> Drunk with fatigue; deaf even to the hoots
> Of gas-shells dropping softly behind.
>
> —Wilfred Owen

By the end of the Third Battle of Ypres in 1917, nearly half a million soldiers were dead, and the town was a ruinous heap.

A century later, visiting the rebuilt Ypres is a solemn occasion, and on his visit in 2014, Scott found that sense of solemnity reinforced by a network of outdoor speakers piping classical music throughout the town center. It was a nice touch, and all the choices were conventionally tasteful . . . that is, until an unexpectedly modern bass line intruded. It wasn't easy to identify the song at first, but the lyrics soon left no doubt. Whoever was behind the music in Ypres had chosen to play the 2006 hit "SexyBack," by Justin Timberlake, throughout the town.

Maybe it had been intentional. Ypres had indeed brought the sexy back to its medieval streets, rebuilding brick by gorgeous brick after the Great War. Nonetheless, in light of all the classical songs, it seemed an odd choice. So when he visited the tourist office for a map of the surrounding battlefield memorials, Scott innocently asked the nice Flemish lady behind the desk whether she had any favorite music for the town speakers.

"Oh, no," she said. "Actually, we just use Spotify."

Even the best recommender systems make a bad suggestion once in a while.

2

THE CANDLESTICK MAKER

What does measuring the size of the universe have to do with saving honeybees? The answer lies in how computers learn to recognize patterns in data, and how they use those patterns to make strikingly intelligent predictions.

IN 2017, OFFICIALS in Beijing realized they had a problem. A significant criminal element was operating in their midst.

Luckily, a careful analysis of these criminals' misdeeds revealed a pattern. Their main target seemed to be the Temple of Heaven Park, home to one of the great architectural masterpieces from the Ming dynasty, and a place of great spiritual significance for the Chinese people. They followed a very specific mode of operation, arriving at the park early, blending in with the senior citizens who gathered there to exercise or sing songs, and waiting patiently until they could be sure their target's coffers were full. Come midmorning, they would put their plan into motion, slipping nonchalantly into a nearby public toilet and lingering inside for a minute or two, so that the timing wouldn't raise suspicion. Then they would strike like lightning, grabbing every roll of toilet paper they could find, stuffing them in a backpack, and walking away from the toilet as if nothing had happened. They were most exposed during those first few steps back into the light. But once they hit the crowd, they were home free.

These thieves had become exceptionally skilled and daring. They were stealing vast quantities of toilet paper, and the Beijing authorities set out to catch them.

The first step was to install automated toilet paper dispensers in every public toilet near the Temple of Heaven, so that each person would receive exactly 60 centimeters of paper, or six squares. Yet it soon became clear that if these thieves couldn't steal toilet paper by the roll, they were willing to steal it six squares at a time. They simply made a loop through the park, collecting their ration of paper from every toilet along the way, before arriving back where they started. Then they repeated the loop again and again, like some kind of klepto-scatalogical carousel ride, until all the toilet paper was theirs. The whole operation took a lot longer than in those free and easy days of unguarded rolls, but the result was just as ruinous for the city's toilet paper budget.

Clearly these thieves would not be stopped by a six-square dispenser. The authorities briefly considered the obvious human-intelligence approach: hiring bouncers to stand guard over the toilets. This being 2017, however, they decided to adopt the artificial-intelligence approach instead: they installed cameras and face-recognition software, powered by something called a "deep learning" algorithm, on every public toilet in the park.

Today, if you hope to use a toilet near the Temple of Heaven, you must: (1) take off your hat, glasses, Guy Fawkes mask, etc.; and (2) stare into the camera, which, thank goodness for small dignities, is on the outside. If the software detects that your face has shown up at a nearby toilet in the last nine minutes, then sorry: no six squares for you.

Equipping toilets with artificial intelligence may seem like an extreme, or perhaps just extremely creepy, solution to the problem of toilet paper theft. Many people have raised privacy concerns—and predictably, many logistical problems have arisen, from long lines to broken cameras to mistaken identities. Our goal in relating this example is certainly not to endorse Beijing's approach but to highlight a simple fact of life: AI-based pattern recognition is everywhere these days—even on toilets. So if you want to understand the modern world, it helps to understand how these systems work and why they depend so strongly on data.

Input/Output: How Machines Recognize Patterns

People are brilliant at recognizing patterns. From a very early age, for example, we learn to match faces with people, and much of our subsequent education is about learning to apply the right patterns:

- To speak, you match a sound with the right meaning.
- To read, you match a string of written symbols with the right word.
- To follow etiquette, you match a social cue with the right behavior.
- To practice medicine, you match symptoms with the right diagnosis and cure.
- To be a data scientist, you match a data set with the right way to analyze it.

Whatever the field of knowledge, being smart means knowing lots of patterns—knowing how to match an input with the appropriate output.

People aren't the only ones who can recognize patterns. For example, one of your authors (Scott) has a sweet-hearted little tuxedo cat who hates road trips. The cat has learned that when his owners start packing, he's going to have to spend some time in the car. Now anytime someone removes a duffel bag from the closet, Markov immediately hides under the bed, just in case.

Nowadays computers can learn patterns, too, just like cats and people. You may recall the story of Makoto Koike, who built a cucumber sorter that exploited the pattern-recognition capabilities of AI. There the input was an image, the output was a decision to sort the cucumber into one of nine different classes, and the pattern was the relationship between the cucumber's visual features and its class. In AI, that's called "image classification," and it's used everywhere—by toilets in Beijing; by Facebook, to identify your friends in untagged photos; and by CERN, the huge physics lab in Geneva, to detect collisions between subatomic particles in images from high-energy physics experiments. But the input doesn't have to be an image. Ultimately, computers are agnostic about the type of input you give them, because to a computer, it's all just numbers. The input might be a sound wave (for interpreting a request to a digital home assistant), a sequence of genes (for predicting someone's susceptibility to a disease), or an English phrase (for translation into Spanish). As the table below indicates, anything you can represent as a set of numbers can be used as an input in one of these pattern-recognition systems. As we'll discuss later, though, sometimes it's obvious how to represent an input as a number, and sometimes it isn't.

Input	Output
	Geolocation: "Rome, Italy"
	Speech to text: "Aus-tin brek-fast tah-koze."
© National Cancer Institute; photo by Renee Comet	Image classification: "Hot dog" / "Not hot dog"
68°F/20°C, 70% humidity, mostly sunny	Numerical prediction: "Power consumption in London will be 25,500 megawatt-hours."
"Buenos días!"	Translation: "Good morning!"
"To be, or not to be . . ."	Author attribution: "Shakespeare."

In this chapter, you'll learn the two key ideas behind how these pattern-recognition systems work:

1. In AI, a "pattern" is a prediction rule that maps an input to an expected output.
2. "Learning a pattern" means fitting a good prediction rule to a data set.

There's a bit of math involved here, but have no fear: these ideas are actually quite simple and elegant once you get to know them, and we're going to spend the rest of the chapter helping you do just that.

Let's first see a quick example of the kind of thing we mean. You may have heard the following rule from a website or an exercise guru: to estimate your maximum heart rate, subtract your age from 220. This rule can be expressed as an equation: MHR = 220 – Age. This equation provides a mathematical description of a pattern in a data set: maximum heart rate (the output) tends to get slower with age (the input). It also provides you with a way to make predictions. For example, if you're 35 years old, you predict your maximum heart rate by plugging Age = 35 into the equation, which yields MHR = 220 – 35, or 185 beats per minute.

A prediction rule in AI is exactly like that: an equation that describes a pattern in the relationship between the input and output. Once you've used a data set to find a good prediction rule, then any time you encounter a new input, you can plug it in to predict the corresponding output—just like you can plug your age into the equation "MHR = 220 – Age" and read off a prediction for your maximum heart rate.

Here's a bit of lingo. In AI, prediction rules are often referred to as "models"—for example, a "face-recognition model" for taking an input of an image and outputting a person's identity, or a "machine-translation model" for taking an English sentence as an input and outputting a Spanish translation. The process of using data to find a good prediction rule is often called "training the model." We like the word "training" here, because it evokes the incremental benefits in fitness that accrue with each new gym workout—or in the case of a model in AI, the incremental improvements in prediction that accrue with each new data point. If we can't hit the gym ourselves, then at least our models can.

But this raises a whole lot of questions. What does it mean to "train a model" on a data set? What makes one model better than another? How would you explain that to a computer—how would you teach an algorithm to find the right pattern in a data set? Moreover, don't computers "think" only in terms of numbers? How does this all work when the input is something complex, like an image or a sound wave, rather than a single number, like a person's age? Perhaps most pressing for those who seek a deeper understanding of AI: Where did this idea—training models with data—come from in the first place? And how did this idea come to play such a central yet invisible role in our lives, sitting there in the background behind everything from social media to cancer therapy, from farming cucumbers to translating Spanish, and from toilets to power grids?

A Stellar Discovery

To answer these questions, we'll start by walking you through an extended example of how a prediction rule can represent a pattern. This pattern is exactly like the ones that come up all the time in AI, and it's much more interesting than the age-versus-heart-rate pattern. This pattern, in fact, led to one of the greatest intellectual triumphs of all time, by helping scientists answer a question they'd been asking for millennia: How big is the universe?

Any curious person today can open up a web browser and find thousands of spine-tingling images from the Hubble Space Telescope: colliding galaxies, remnants of exploded stars, distant quasars with the energy of a million suns. Astronomers of only a century ago, however, would hardly recognize these wonders. For them, the universe was a much smaller place. Back in 1924, enlightened scientific opinion held that our galaxy, the Milky Way, was the only galaxy in the universe—and beyond its horizon, only a void. It wasn't until the early twentieth century that people finally learned the awesome truth: we inhabit a vast universe of a trillion galaxies or more.

We will focus here on three essential features in the story of this great discovery:

1. an unexplained smudge of light in the night sky, visible even to the ancients;
2. a centuries-old mathematical principle for pattern recognition that today powers our most sophisticated AI systems; and
3. a little-known astronomer of the early twentieth century named Henrietta Leavitt who, by using that principle, taught us how to measure the size of the universe.

When you see how these three strands fit together, you'll come away with a far richer understanding of how machines learn patterns in the world around them, and how they use those patterns to make strikingly accurate predictions—whether that involves classifying cucumbers, recognizing your friends in photos, or wiping out toilet paper theft in Beijing.

A "Misty Smear" in the Northern Sky

More than a thousand years ago, keen observers first took notice of a few small wisps in the sky—not stars, exactly, more like hazy clouds of light. The biggest one, visible to the naked eye on a dark night, was a glowing patch at the waist of Andromeda, a constellation in the northern sky. In the tenth century AD, the Persian astronomer Abd al-Rahman al-Sufi referred to this object as a "misty smear."[1] Al-Sufi couldn't figure out what it was, and neither could anyone else. When the telescope came around in the 1600s, the mystery of the "Great Andromeda Nebula" only deepened, as astronomers began to discover even more little smears just like it—many of them, again like Andromeda, with the unmistakable shape of a spiral.

By the 1800s, astronomers were calling them "spiral nebulae," after the Latin word for "mist." These nebulae cried out for an explanation. Were they newborn stars? Were they clouds of luminous gas in the outer reaches of the Milky Way? Or was each one, as a few people claimed, a distant galaxy just like our own?[2]

Figure 2.1. A modern image of Andromeda from NASA's Galaxy Evolution Explorer. Courtesy NASA/JPL-Caltech

This last interpretation—that the spiral nebulae were "island universes," the term then used for a galaxy—had been popular for much of the eighteenth and nineteenth centuries. Its most famous champion had been the German philosopher Immanuel Kant. By the early twentieth century, though, the island-universe theory had fallen out of favor. There was no direct evidence for it, so most astronomers had settled on the simpler "one-galaxy" hypothesis. They concluded that the spirals sat on the outskirts of the Milky Way and were probably clouds of newly forming stars. The idea that they were independent galaxies came to be viewed as "grandiose" and "misleading," a notion that one astronomy textbook of the day described as so silly that it "hardly any longer needs discussion."[3]

Yet as telescopes got better and new evidence accumulated, a few astronomers began to wonder whether they'd been too quick to dismiss the old independent-galaxy theory. One point in their favor was the pace at which astronomers were discovering novas: "new stars" that appeared suddenly in the night sky and then gradually faded from view over weeks or months. People had been seeing novas for hundreds of years, but the powerful new telescopes of the early 1900s presented astronomers with a curious fact. The Great Andromeda Nebula seemed to have a surprising number of novas in it—more, in fact, than the rest of the galaxy combined. If Andromeda were just a cloud of dust on the outer fringe of the Milky Way, how could this be? Why would one tiny corner of our galaxy be so incomparably rich in novas?

Then there was the issue of how fast Andromeda was moving. In 1913, the astronomer Vesto Slipher had painstakingly measured its speed using a spectrometer, a cosmic "radar gun" that works by exploiting the Doppler effect—the same principle that causes the siren of an ambulance to sound higher as it approaches you, and then lower as it recedes into the distance. Slipher's results were so astonishing that he hardly believed them himself: Andromeda was moving relative to Earth at a rate of 300 kilometers *per second,* about 20 times faster than anything else in the Milky Way. More shocking still was the fact that most of the other spiral nebulae were moving even faster than Andromeda—many as fast as 1,000 kilometers per second. For many astronomers, Slipher's results put the matter to rest: the spirals were moving too fast to be inside our own galaxy.[4]

Skeptics of the independent-galaxy theory, however, had a ready rejoinder. If the Andromeda Nebula were a galaxy the size of our own, two seemingly impossible conclusions would follow. For one thing, Andromeda would have to be millions of light-years away, or else it would be much brighter in the night sky. And if *that* were true, then each of Andromeda's novas would have to be burning with the energy of millions of suns, or else we'd never see them from so far away. In hindsight, we now know that both "impossibilities" are actually true. But in the eyes of many early twentieth-century astronomers, they reduced the independent-galaxy theory to an absurdity.

What, therefore, were astronomers to make of all those nebulae? Were they small or large? Dust clouds in our own galaxy, or entire galaxies of their own? No one had decisive evidence either way—which left astronomers in a horrible muddle, for upon these questions turned the answer to an even deeper one: How big, really, is the universe? Copernicus had humbled us once, disproving that Earth was the center of creation. Galileo had humbled us a second time, showing that the Milky Way was a huge throng of stars just like our own. Were we soon to be humbled a third time, by learning that our galaxy wasn't alone? It was astronomy's "Great Debate," and it raged on throughout the 1910s and early 1920s. And the only reason it raged was because no one could answer a simple question: How far away is the Great Andromeda Nebula?

How Can You Measure the Stars?

Imagine driving down an unlit country road on a dark night. You crest a hill, and a light winks into view up ahead. How far away is it? Are you seeing the dim porch light of a house a few hundred feet away? The headlights of another car a mile down the road? Or perhaps something farther off but much brighter, like the glow of a small town 10 miles away in the valley?

You are now facing the fundamental problem of astronomy. Your eyes can only tell you how bright an object *seems*, not how bright it really is at the source. Venus, for example, seems like the brightest thing in the night sky other than the moon, but only because it's so close. The star Alpha Centauri, meanwhile, seems 100 times dimmer than Venus, but only because it's 25 trillion miles away. Up close, it's brighter than our sun.

A telescope has the same problem. It can measure a star's *apparent* brightness, or how bright it seems to us here on Earth, but not its *true* brightness, or how much light it's actually giving off. This leaves astronomers asking the same question of every dot of light in the sky: Is it dim and close, or bright and far away?

You might ask: If that's true, then how do we know that Alpha Centauri is 25 trillion miles away? The answer is that astronomers exploit a useful pattern called "parallax." You can see parallax for yourself by playing left-eye/right-eye: raise your index finger, or perhaps a different finger if you're cursing us for all the math, and hold it a few inches in front of your nose. Look at it first with one eye closed, then the other. Keep switching back and forth between eyes. You should notice that your finger seems to move from one eye to the other. That apparent movement is parallax.

Now here's the pattern: the farther you hold your finger from your nose, the less it appears to move as you switch back and forth between eyes. You can describe that pattern mathematically, as an equation that relates your finger's apparent motion, or parallax, to its distance from your nose. The equation says that as parallax gets smaller, distance gets bigger: distance = 1/parallax. This can be derived using trigonometry; we'll spare you the details, because the point is that it's just a case of "output = function of input" all over again, like the prediction rule that relates your maximum heart rate to your age.

Astronomers measure the distance to nearby stars by playing the same left-eye/right-eye game. Specifically, they use two telescope images of the star taken half a year apart. This allows Earth to complete half an orbit around the sun, maximizing the separation between the astronomers' left and right "eyes." They compare these two images to measure the star's parallax, which they plug into the equation above to yield a prediction for the star's distance.

But the big downside to the parallax/distance pattern is that, as a cosmic measuring tape, it doesn't stretch very far. If an object is farther away than about 300 light-years, its parallax will be too small to measure reliably—and 300 light-years is barely an inch by galactic standards. Even in the early 1900s, as the Great Debate over the spiral nebulae was at its rowdiest, everyone accepted that the Milky Way was at least tens of thousands of light-years across, if not more. Therefore, both sides recognized that no matter who was right, the distance to Andromeda was too far to measure using parallax. Astronomers were desperate for a better way to measure distance, but nobody had one.

Nobody had one, that is, until a little-known astronomer named Henrietta Leavitt made a wonderful discovery. Leavitt found a new prediction rule that would allow astronomers to measure distances over millions of light-years, much farther than they had ever thought possible. She didn't find the rule using trigonometry, the way the parallax rule had been found. Instead, she found it using data, by applying the same principle that Google, Apple, and Facebook use today to build their pattern-recognition systems.

Leavitt's Great Discovery

Henrietta Leavitt ended up as an astronomer almost by chance. Born into a large family in Lancaster, Massachusetts, in 1868, she entered Radcliffe College in 1888 to study the humanities. It was only in her senior year that she even took an astronomy course. Luckily for science, she loved it so much that she ended up staying on after completing her degree, to take graduate-level courses and to volunteer at the Harvard College Observatory. Her outstanding abilities soon drew the notice of the observatory's director, Edward C. Pickering, who asked Leavitt to join the "Harvard Computers," a team of math prodigies—all women—hired to analyze data from telescopes. Long before a "computer" was a device, it meant a person who did calculations.[5]

Leavitt's main role was to estimate and catalog the brightness of stars for Harvard's massive ongoing "sky survey," which required her to compare the sizes of tiny spots of light across thousands of archival images from the world's great telescopes. It was repetitive, painstaking work; humans haven't always been so lucky as to have algorithms that can extract patterns from images automatically.

But one thing kept Leavitt on her toes during all this drudgery, and that was the search for *pulsating stars*. A pulsating star is one whose brightness changes over time in a strikingly regular way: it oscillates from bright to dim to bright again, over and over like clockwork. (See Figure 2.2.) We now know that these pulsating stars are thousands of times larger than our own sun, and that they fluctuate in brightness because their stellar atmospheres are repeatedly swelling and shrinking, like your lungs as you breathe. In Leavitt's day, though, precious little was known about these curious stars. Astronomers were fascinated by them, and Leavitt had standing instructions to keep an eye out for every one she could find.

To do this, she collected images of a single star taken across many different nights. She pored over them with a magnifying glass and a tiny ruler to check for the telltale sign of a pulsating star: whether its spot of light got bigger and smaller repeatedly over time. She did this image by image, star by star, for years on end, and she ended up finding 1,777 pulsating stars previously unknown to science.

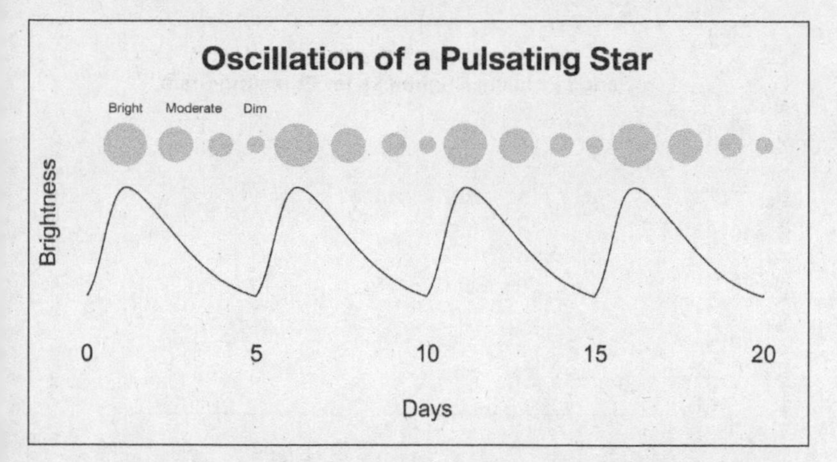

Figure 2.2. The oscillation in brightness of a pulsating star. This particular star completes a cycle from bright to dim to bright again every 5.4 days. Leavitt discovered that a pulsating star's period was related to its brightness: brighter pulsating stars oscillate more slowly than dim ones, in a mathematically predictable way.

By 1912, Leavitt had zeroed in on a group of 25 pulsating stars in the Small Magellanic Cloud.* Because these stars were all part of the same cluster, Leavitt felt safe in assuming that they were all about the same distance from Earth. Therefore, if a star *seemed* brighter, it actually *was* brighter at the source. For each star, she tabulated two data points. First, there was its period of pulsation, or how long the star took to complete one full cycle from bright to dim to bright again. Each star had a specific period, ranging from 1.25 days all the way up to 127 days. Second, there was the star's brightness, or how much light it was giving off.

Leavitt then plotted her data. In Figure 2.3, we've made our own version of this plot using her original data. Each point is one of Leavitt's 25 pulsating stars. The horizontal (X) coordinate represents the star's period of pulsation, and the vertical (Y) coordinate represents its brightness. Plots like these are great for revealing patterns in data sets—and Leavitt's revealed an extraordinary pattern involving a pulsating star's

* These were later discovered to be a special class of pulsating stars called "Cepheid variables."

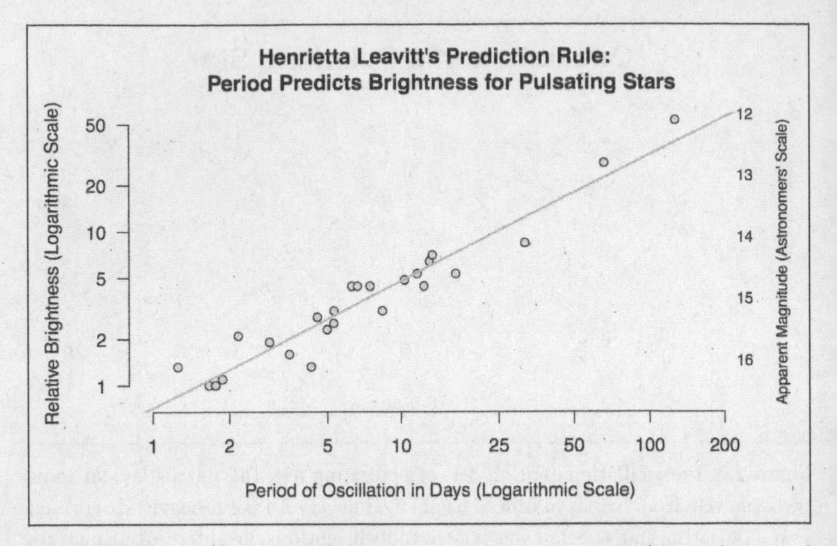

Figure 2.3. Henrietta Leavitt's 1912 data on 25 pulsating stars. This mathematical pattern—a straight line that relates period of oscillation with brightness—allowed astronomers to measure cosmic distances over previously unimaginable scales.

brightness and its period. Her data points, it turned out, fell almost perfectly along a straight line. The dimmest stars in her data set had periods measured in days, while the brightest stars had periods measured in months. The longer the period, the brighter the star. It was a pattern so regular that you could describe it with an equation: a straight line drawn right through the middle of the data points.

It turned out to be one of the most important straight lines in the history of science. To understand why, imagine again that you're back on that dark country road, where you see a light in the distance, but you don't know how far away it is. Now imagine that someone gives you a clue, by telling you exactly how bright the light is at the source. From this clue, you can work out the distance to the light: if you're told that the light is a single 60-watt bulb, you know it must be nearby, but if you're told that it's the light from a whole town, you know it must be far away. The general principle is that if you can measure the object's *apparent* brightness, and if someone tells you that object's *true* brightness—how much light it's actually giving off—then you can work backward, using the laws of physics, to figure out how far away the object is. That working-backward process is mathematically tedious but conceptually simple. The real clue that unlocks the mystery of distance is *knowledge of an object's true brightness.*

Leavitt's discovery provided astronomers with exactly that kind of clue. They could point their telescopes at a pulsating star and measure both its apparent brightness and its period. Then they could take the star's period, plug it into Leavitt's equation, and read off the corresponding prediction for its *true* brightness.† This immediately yielded the star's distance—and therefore the distance to any *other* stars in its vicinity. In the parlance of astronomy, Leavitt had discovered that pulsating stars were "standard candles": objects of known brightness whose distance could be reliably measured. In the parlance of AI, she had discovered a prediction rule. Her equation was just "output = function of input" all over again.

Leavitt published her results in 1912, in a paper only three pages long. Her colleagues immediately recognized that her discovery provided the cosmic measuring tape they'd been seeking so eagerly, and they put it to use as quickly as their instruments would allow.

The first major result came from an astronomer named Harlow Shapley. He measured the period of several pulsating stars in the Milky Way, applied Leavitt's prediction rule to work out their true brightness, and then used the result to calculate their distance. These stars turned out to be surprisingly far away. Shapley's findings implied that our galaxy was at least 100,000 light-years across, much larger than anyone had imagined—and, in an echo of Copernicus, that our own sun was nowhere near the galactic center.[6]

---

† This is a slight oversimplification. Leavitt's line allowed you to work out the true brightness of a pulsating star, *relative to* the true brightness of any other pulsating star. So the technically correct statement is this: once you know the true brightness of one pulsating star—just one—then from Leavitt's pattern, you can immediately work out the true brightness of any other pulsating star in the universe. In the parlance of astronomy, Leavitt's pattern still had to be "calibrated," by estimating the true brightness of a single pulsating star by some other means. It took astronomers several years to figure out how to do this, which is why Leavitt's pattern couldn't be used immediately to measure stellar distances. But we'll leave that part of the story to others; see, for example, Marcia Bartusiak's *The Day We Found the Universe* (New York: Vintage Books, 2010), chapter 8.

But the real blockbuster came from another astronomer, now one of the most famous scientists of all time: Edwin Hubble.

In 1919, Hubble started a job at the Mount Wilson Observatory in Pasadena, California, arriving just in time for the opening of the new 100-inch Hooker Telescope. With Leavitt's prediction rule at the front of his mind, he began searching for pulsating stars in spiral nebulae—and since he had the largest telescope in the world at his disposal, he stood a very good chance of finding them. Each pulsating star would serve as a beacon, a standard candle that Hubble could use to calculate the distance to its host spiral.

It took years, but Hubble's slow, careful search paid off. In October of 1923, Hubble finally experienced his eureka moment: he found a pulsating star in Andromeda, the same "misty smear" that had caught the attention of Abd al-Rahman al-Sufi over 1,000 years before and had flummoxed every astronomer since. He measured the star's apparent brightness, and he computed its period at 31.4 days. He plugged this value into Henrietta Leavitt's prediction rule to get its actual brightness. He then worked backward, using both the true and the apparent brightness to calculate a distance to Andromeda.

The result was a revelation. The Great Andromeda Nebula was over a million light-years from Earth—far outside the Milky Way. Andromeda was therefore *enormous*, since we could see it from Earth at such a vast distance. It could only be one thing: a galaxy unto itself. At a stroke, Hubble had settled a question about our place in the cosmos that had been open for more than a millennium.

Hubble would later go on to use the pulsating-star technique to discover many more galaxies—or, as he put it, "Whole worlds, each of them a mighty universe, are strewn all over the sky . . . like the proverbial grains of sand on the beach."[7] Yet it was that first pulsating star that he found in Andromeda—known today as "Hubble variable 1," or V1—that went down in history. Decades later, in 1990, when the space shuttle Discovery carried the Hubble Space Telescope into low Earth orbit, it also carried something whose value was entirely sentimental: a copy of Hubble's original photograph of V1 from 1923.[8] It was a photograph that made Hubble a household name and that changed the course of astronomy forever. But it was also a photograph whose significance Hubble could have seen only by standing on the shoulders of Henrietta Leavitt—for she was the one who showed Hubble, and everyone else, how to measure the size of the universe.

Fitting Prediction Rules to Data

We'll return to the story of Henrietta Leavitt at the very end of the chapter. For now, however, let's keep her great discovery in mind as we revisit the two key ideas about pattern recognition that we mentioned at the beginning of the chapter.

1. In AI, a "pattern" is a prediction rule that maps an input to an output.
2. "Learning a pattern" means fitting a good prediction rule to a data set.

We hope that Henrietta Leavitt's pulsating stars have taught you the value of a good prediction rule for describing a pattern. But we also expect that you still have some questions. For example, what makes one prediction rule better than another? And how can something as literal-minded as a computer learn the right prediction rule?

In AI, the criterion for evaluating prediction rules is simple: How big are the errors the rule makes, on average? No prediction rule can be perfect, mapping every input to exactly the right output; all rules make errors. But the smaller the average error, the better the rule.

To understand this, let's take a second look at Henrietta Leavitt's prediction rule for pulsating stars. In the left panel of Figure 2.4, you see Leavitt's data, together with her original equation: the straight line that relates a pulsating star's brightness to its period. The scale of brightness here is the one astronomers use, called "magnitude"; for historical reasons, astronomers keep score like golfers, where smaller numbers mean brighter stars.

Focus on the star we've highlighted with an arrow, with a period of about two days. We can measure how much Leavitt's rule misses here

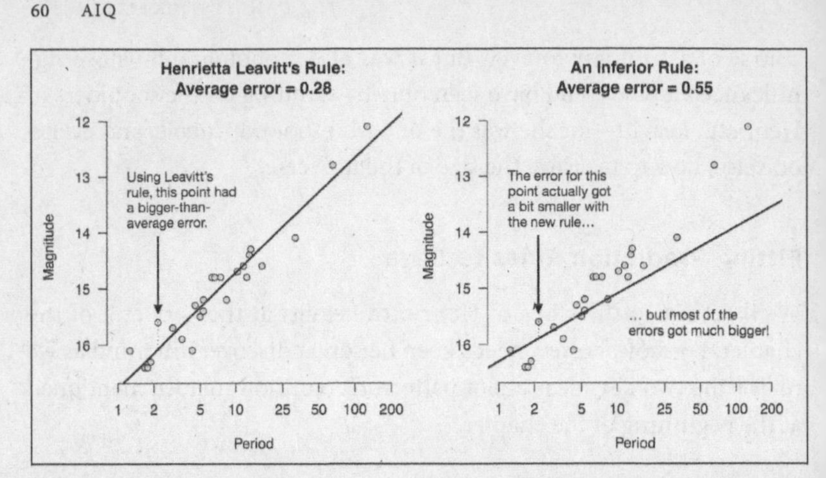

Figure 2.4. Leavitt's original equation (left) versus a modified equation (right) that doesn't fit the data as well.

by calculating the vertical distance between the point and the line; this distance is called the "residual" or the "reconstruction error." For this star, Leavitt's rule predicted that the stellar magnitude should be about 16.1, whereas in reality it was about 15.6—a reconstruction error of 0.5 units.

Now consider a slightly different rule, like the one in the right panel. Here we've tweaked Leavitt's line by making it a bit less steep. For the one star we've highlighted, the error actually got smaller. But for most of the other stars, the errors got bigger. Leavitt's rule is therefore better than ours—on average, it's nearly twice as accurate.

In fact, it turns out that Leavitt's prediction rule is the best rule: among all straight lines, it is the one with the smallest possible average reconstruction error. You can tweak the line all you want, and some of the individual errors might get smaller. But your tweaks, like ours in the right panel, are guaranteed to make the average error get bigger. We know this because Leavitt followed a mathematical recipe called "the principle of least squares" to fit a prediction rule to her data. First published by the French mathematician Adrien-Marie Legendre in 1805, this principle provided an explicit formula for fitting the "optimal" straight line to a data set—that is, the line resulting in the smallest possible average reconstruction error.‡ Scientists have been using this formula ever since.[9] Moreover, the same basic principle that Legendre articulated over two centuries ago is still being used today, to build some of the world's most sophisticated artificial-intelligence systems. Prediction rules in AI are just fancier versions of the prediction rule discovered by Henrietta Leavitt: they're equations that map inputs to outputs, and they're chosen to minimize the average reconstruction error on a data set, just as Legendre suggested more than two centuries ago.§

Before discussing this idea in detail, we'll give you three quick examples, all of which you can use directly from your phone. First, take image-recognition software, like the kind that identifies your friends in the photos you upload to Facebook. Image recognition is just a prediction rule: the input is an image of a person's face, and the output is that person's identity. The mapping from inputs to outputs is a complicated equation that describes a complicated pattern in the training data: which facial features tend to go with which names in previously uploaded photos.

Second, consider Google Translate. This is also just a prediction rule: it maps input phrases in one language (say, English) to output phrases in another language (say, Spanish). The underlying model is, again, a complicated equation that describes a complicated pattern: which English phrases tend to go with which Spanish phrases across a huge database of sentences rendered side by side in both languages.

Finally, consider a new smartphone app developed by Dr. Elina Berglund Scherwitzl, a Swedish physicist who helped discover the Higgs boson—and who's now well into her second career as an entrepreneur, having invented a new AI-based contraceptive technology. Berglund Scherwitzl had long sought an alternative to hormonal contraception,

---

‡ Technical note: if you've ever taken a course in calculus, you might remember that you learned how to minimize functions. Legendre did exactly this kind of thing in order to find the prediction rule that minimized the average error. Actually, Legendre's solution minimizes the average *squared* error (hence "least squares"). This is an important technical point, but not at all important for understanding the basic idea.

§ This is a slight oversimplification. We also have to worry about something called "overfitting," which we'll discuss in a few pages.

yet she had never managed to find an option she liked. In that problem, she saw an opportunity. She and her husband, Raoul Scherwitzl, quit their jobs as physicists and set out to use their data-science skills to build a new variation on an old idea: the "rhythm method," which involves tracking your menstrual history to predict when you're most likely to be ovulating.

The problem with the traditional rhythm method is that to use it successfully, you need an uncommonly meticulous approach to record-keeping. Berglund Scherwitzl's version relies on something more reliable: body temperature, whose monthly cycle correlates strongly with fertility. To use the method, you enter two pieces of information into a smartphone app, Natural Cycles: your daily body temperature and your date of menstruation. Over time, as you give the app more training data, it fits a prediction rule customized to the patterns in your own cycle. The input is your temperature, while the output is a prediction about your fertility that day, in the form of a little traffic light on your smartphone. (Green means go.) Apps that help you track your menstrual cycle are very popular, but this is the first one to have been certified as an effective contraceptive method by regulators in the European Union. In clinical trials, the app was shown to be roughly as effective as the pill at preventing pregnancy under typical usage patterns.* As of mid-2017, it was helping over 300,000 subscribers take control of their reproductive choices, using AI.

Beyond Straight Lines

At this point, you might be wondering something. We've explained that in AI, recognizing a pattern means fitting an equation to data. And we've also explained that this idea goes back to 1805. What, then, accounts for the recent revolution? Why have all these pattern-recognition systems, from face detection to machine translation to AI-based birth control, arrived on the scene only within the last few years?

---

* The pill is much more effective under "perfect use," or use as the manufacturer intended, with only a 0.3% failure rate over a single year.

Here's the basic issue: the patterns to be found in massive databases of images, text, and video are radically more complicated than any pattern you can visualize with a scatter plot, like Henrietta Leavitt's plot of pulsating stars. And these complicated patterns must be described with complicated equations—much more complicated, at least, than the equation of a straight line. Such complicated equations are highly demanding: as we'll explain, you need a lot of computational horsepower to work with them, and you need lots of data to be able to estimate them reliably. Only recently has our technology made this feasible and cheap.

The big breakthrough in AI involves the use of *neural networks* for estimating prediction rules from data. The term "neural network" sounds awfully brainlike, but that's nothing more than a brilliant piece of marketing. In reality, a neural network is just a very complicated equation that's capable of describing very complicated patterns in data—that is, very complicated mappings from inputs to outputs. The reason we use neural networks is not because they replicate what human brains do but because they work incredibly well across an astonishing range of prediction tasks, from language to images to video.

Let's take a closer look at the four factors driving this breakthrough.

Factor 1: Massive Models

We used to build prediction rules to describe simple patterns using small models—sort of like mining data with the equivalent of picks and spades. Today, we describe complicated patterns using massive models—more like one of those giant mining trucks with tires the size of a small house. It's the same idea, just with a much bigger shovel.

To understand what we mean by a "massive" model, you have to understand the concept of a *parameter*. A parameter is a number in your equation that you're free to choose in order to yield the best possible fit to the data. Small models have a few parameters, while massive models have lots of parameters. For example, you might remember this equation from earlier: Max Heart Rate = 220 – Age. This is a small

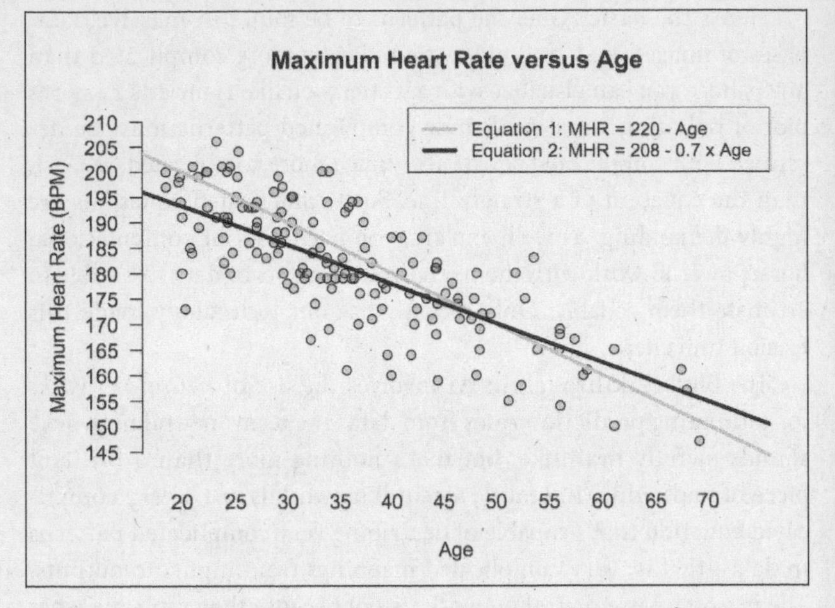

Figure 2.5. Two equations for predicting your maximum heart rate from your age. The black line has one parameter; the gray line has two parameters, and it fits the data better as a result.

model, because the equation has just one parameter: the baseline of 220, from which you subtract your age. We could have chosen a baseline of 210, or 230, or anything—but the choice of 220 fits the data best.

But there's actually a slightly bigger model that works even better: Max Heart Rate = 208 – 0.7 × Age. In words, multiply your age by 0.7 and subtract the result from 208 to predict your maximum heart rate. The previous rule had only one parameter, while this new rule has two: the baseline of 208 and the "age multiplier" of 0.7, both of which can be tuned to fit the data. You can see both of these rules in the figure above, which shows a scatter plot of 151 adults whose maximum heart rates were measured in a laboratory. The plot shows both prediction rules: the old "220 – Age" rule in gray, and the new "208 – 0.7 × Age" rule in black. Exercise scientists prefer the black line; having two parameters rather than one gives them extra flexibility to tune the equation so that it fits the data as well as possible.[†] Some people still repeat the old "subtract your age from 220" rule because it's simpler—one parameter rather than two. But you pay a price for simplicity. Its prediction about your maximum heart rate won't be as accurate as the prediction from the two-parameter model, at least on average.

Now let's see what a prediction rule with *three* parameters looks like. Suppose that you're a data scientist at the online real-estate marketplace Zillow, and that you have to build a rule for predicting the price of a house.[‡] You might start with two obvious features of a house, like the square footage and the number of bathrooms, together with "multipliers" for each feature. For example:

Price = 10,000 + 125 × (Square Footage) + 26,000 × (Number of Bathrooms)

In words, this says that to predict the price of a house, you should follow three steps:

1. Multiply the square footage of the house by 125 (parameter 1).
2. Multiply the number of bathrooms by 26,000 (parameter 2).
3. Add the results from 1 and 2 to the baseline of 10,000 (parameter 3). This yields the predicted price.

But why stop at only two feature multipliers? Houses have lots of other features that affect their price—for example, the distance to the city center, the size of the yard, the age of the roof, and the number of fireplaces. Using the principle of least squares that Legendre articulated in 1805, data scientists can easily fit an equation that incorporates all of these features, and hundreds more. That's basically how Zillow builds its prediction rule for the price of a house. Every feature gets its own multiplier, and more important features end up getting bigger multipliers, because the data shows that they have a bigger effect on the price.

† The "MHR = 208 – 0.7 × Age" rule is the least-squares fit to the data using Legendre's formula from 1805.

‡ Zillow is known as Zoopla in the U.K.

Of course, if you try to write out such a prediction rule in words—"add this," "multiply this," like we did for the two-feature rule above—it starts to look like the IRS forms they hand out in hell. But a computer just churns through all the calculations with no problem, even for models with hundreds of parameters.

In AI, however, we dream even bigger than that, fitting models with *far* more than a few hundred parameters. Take, for example, a model for labeling images. To a machine, an image is just pixels, and pixels are just numbers: color intensities from 0 to 100%. An uncompressed 1 megapixel image, for example, has 3 million numbers associated with it: a red, green, and blue intensity for each of 1 million pixels. That's 3 million features right there. And you need a lot of parameters to be able to make the best use of 3 million features—especially if you're going to combine those features in interesting ways, rather than just give each one a single multiplier, the way we did in our imaginary Zillow model above.[§]

This is where neural networks come in. In 2014, for example, engineers at Google published a paper about a neural-network model—nicknamed "Inception," after the Leonardo DiCaprio film—that could automatically recognize and label an image. And it was shockingly effective. Older image-recognition models might have been able to tell you whether a given photo was a dog or a not-dog; Inception could tell you whether a given dog was a Siberian Husky or an Alaskan Malamute. The model involved 388,736 parameters, and using it to make a prediction required 1.5 billion arithmetical operations—1.5 billion little steps of "add this" or "multiply this"—*for a single input image.* That's a long IRS form. It's a good thing that an Nvidia graphics card in 2018 can do 1.5 billion calculations in less than 0.0001 seconds.

Factor 2: Massive Data

But there's a caveat: to fit a massive model, you need a massive data set.

A model like Google's Inception, with 388,736 parameters, tends to

§ In AI-speak, this model would be "nonlinear," as opposed to our imaginary Zillow model with a single multiplier for every feature.

blow the minds of old-school scientists and engineers, who regard such massive models with contempt. The great mathematician John von Neumann, for example, once famously criticized a complicated model with the following enigmatic quip: "With four parameters I can fit an elephant, and with five I can make him wiggle his trunk." Von Neumann meant that a model with lots of parameters is in danger of "overfitting," which happens when a model just memorizes the random noise in the training data rather than learns the underlying pattern. An overfit model may describe the past with perfect accuracy, yet still be bad at predicting the future.

If you want to understand overfitting, look no further than all those political pundits you see on TV who get paid to come up with absurd nuggets of "wisdom" about presidential elections—for example, "No Democratic incumbent without combat experience has beaten someone whose first name is worth more in Scrabble." This prediction rule actually *did* hold for 208 years of American history, until Bill Clinton beat Bob Dole in 1996.* But it never had any value for predicting the future. The rule is a classic example of overfitting: retrospectively sifting through thousands of complicated details about past elections and cherry-picking the *one* complicated detail that happened to be true.

So if you're fitting a prediction rule to data, how do you avoid overfitting? There are only two ways. First, you can disallow complicated explanations. Your model can't memorize complicated, nongeneralizable facts about the past if you force it to ignore *all* facts except the simplest ones. This solution works well in the hard sciences. In fact, it's exactly what von Neumann was advocating with his "fit an elephant" line; he was interested in finding simple theories that could explain universal physical laws of matter and energy, rather than contingent earthly particulars like elephants or wiggling trunks. But the "ignore complicated theories" approach doesn't work at all in AI. The patterns we want to find in data—for example, which combinations of pixels are labeled "Husky" and which ones are labeled "Malamute"—actually *are* complicated, earthly, and particular. Small models with two or

* This example is from the brilliant cartoon series *XKCD*: https://xkcd.com/1122/.

three parameters, or even two or three thousand, simply can't explain these patterns accurately.[†]

So that forces us to turn to the second strategy: collect massive amounts of data. Lots of data means lots of experience—and with enough experience, you can rule out the *bad* complicated explanations, leaving only the good complicated explanations behind. This solution doesn't work for presidential elections; there have only been 56 of them, so there is basically no way to tell from the data alone whether a complicated post-hoc explanation of who wins the presidency has any value in predicting the future. But it works brilliantly for models that extract patterns from images, texts, and videos, which we have in abundance.

John von Neumann would surely be amazed at the result. He thought that you could "fit an elephant" with only four parameters, but it turns out you need 388,736 of them—or at least you need that many parameters to *identify* an elephant in the photos from your African safari. There's no magic here, just massive data sets with millions or billions of data points. This lets us use complicated models to describe complicated patterns, without overfitting. And in all fairness to von Neumann, he surely never imagined a world where people uploaded 100 million images a day to Instagram, many of them with helpful labels like #safari or #elephant.

Factor 3: Trial and Error, a Million Times per Second

In the early 1900s, Henrietta Leavitt fit her prediction rule using pencil and paper, by applying Legendre's mathematical formula for optimal straight lines from 1805. Even as late as the early 2000s, most scientists were still fitting prediction rules using minor variations on the same formula. The only difference was that we modern folks had gone soft, allowing a machine to handle the calculations.

But there's no mathematical formula for fitting the prediction rules of today. In fact, there's only one good way to fit a massive model like

† Technical note: designers of models in AI actually do try to make their models simpler, using a mathematical technique called "regularization." This also helps avoid overfitting, and it's really important if you want to fit good prediction rules. We encourage you to read more about overfitting if you're interested in this area.

Google's Inception, and that's incrementally, by trial and error. You start with some initial guess for a prediction rule—for example, that all pictures with gray shapes on green backgrounds are elephants lounging on the savannah. This initial rule will almost surely be terrible. As data starts arriving, however, you refine your rule. For each new data point, you ask two questions: How big an error does my current model make on this data point, and how could I tweak the model to make a smaller error? Modern computers can ask and answer these two questions thousands or even millions of times every second. When a massive data set is subjected to this kind of relentless computational onslaught, it doesn't take long for your prediction rule to improve dramatically—for example, by learning that *some* gray shapes on a green background are elephants, while others are rhinos.

Today, this trial-and-error model-fitting strategy is used everywhere. It's what allows large retailers, for example, to predict what you want to buy online, before you even know you want to buy it. Take Alibaba, the Chinese e-commerce giant that booked $24 billion in revenue in 2017. Like Amazon, Alibaba promises to deliver your stuff fast—so fast that there's no way it could all ship from a central warehouse. Instead, data scientists at Cainiao, Alibaba's logistics arm, have to be masters of AI, predicting exactly what customers will want over the coming days and weeks, so that the company can ship the right items to the right local distribution center long before anyone clicks BUY. They have to do this, moreover, for every product Alibaba sells and for every market it serves, whether it's this neighborhood of Shanghai or that district of Guangzhou. And they do it by trial and error: using massive data sets to train massive models that get a little bit better with every new purchase.

In the AI business, this process of refining a model by intelligent trial and error goes by many different names, like "online learning" and "stochastic gradient descent." We're omitting a lot of details here that are essential for making this strategy succeed, but they're all just minutiae, the kind of thing you learn if you study AI in graduate school. If you just think "trial and error," you're 90% of the way there.

Factor 4: Deep Learning

In addition to the richness of our models, the size of our data sets, and the speed of our computers, there's a fourth major way in which prediction rules have improved dramatically: people have learned how to extract useful information from vastly more complicated inputs. If you've heard the term "deep learning" and wondered what it means, we're about to explain.

We said at the beginning of the chapter that computers are agnostic about the type of input you give them. But that's only sort of true. Henry Ford was famous for saying that customers of Ford Motor Company could buy a car in any color they wanted, as long as it was black. Computers are the same: you can give them an input in any form you want, as long as it's a number. The hard part in most AI applications isn't training the model but answering the question that comes first: How do I represent the input to my model as a set of numbers? Data scientists call this "feature engineering," which just means extracting numerical features from some input that isn't obviously a number, like an image or a string of English words.

Over the last decade, AI experts have gotten dramatically better at automated feature engineering, using a specific kind of prediction rule called a "deep neural network." You learned before that a neural network is just a complicated equation with lots of parameters. A deep neural network is a variation on this idea, where the equation is structured in a way that extracts as much information as possible from a specific kind of input.

Take images. Here, deep neural networks solve a key conceptual challenge in feature engineering: many different possible arrangements of an image's pixels can ultimately mean the same thing. Rotations, translations, changes of color—all of these things can dramatically change the pixels in an image, without changing the content. A red heart symbol, for instance, means the same thing whether you put it on the left of an image or the right, or whether you rotate it a few degrees one way or another. It even means the same thing if you change the color—like in that famous country song from the 1990s by Joe Diffie, when a young farmhand named Billy Bob climbs the local water tower to paint a 10-foot-tall heart, along with a message of love to his sweetheart, Charlene, using the only color of paint he had: John Deere green.[10] As Charlene understands, a heart is a heart, whether it's painted in red or some other, more agricultural color. But a computer that's been programmed to interpret pixels too literally can easily get confused. That's why we need feature engineering: to turn raw pixels into useful, generalizable facts about an image.

Deep neural networks solve this problem in a very clever way. To explain this, we'll return again to Henrietta Leavitt's search for pulsating stars, and her subsequent discovery that they could be used to measure distances to faraway corners of the universe. Let us refresh your memory about what Leavitt had to do here. She had to track a single star across multiple photographs. She had to measure the star's brightness in every photograph, to check whether it waxed and waned in the telltale manner of a pulsating star. If it did, she had to calculate the period of the star, or how long it took to complete one pulsation.

In doing all this, Leavitt had to draw on at least five visual concepts, each at a successively higher level of abstraction.

Level 1: The bright parts of the photo represent light from the sky.

Level 2: A star is a spot of light surrounded by dark.

Level 3: The brightness of a star is the size and intensity of its spot.

Level 4: A pulsating star is a star whose brightness changes in a regular fashion across multiple photographs.

Level 5: The period of a pulsating star is the length of time it takes to go from bright to dim to bright again.

This is what feature engineering looks like. If you follow the hierarchy from level 1 to level 5, out pops a number: the period of a pulsating star, which you can subsequently use as an input in your prediction rule. (Remember, Leavitt's prediction rule for pulsating stars had period as an input and true brightness as an output.)

You might say that Leavitt was using a "five-layer deep neural network" here. She was applying a series of visual concepts, chained together in a five-layer-deep hierarchy, to extract a useful feature from an image.

And that's exactly what a deep neural network does.‡ Each new layer of the hierarchy draws on the lower-level concepts—just like here, where the level 4 concept of "pulsating star" is defined in terms of a level 2 concept (star) and a level 3 concept (brightness). At the very top level of the hierarchy, you end up with a feature—period—that can be used as an input in a prediction rule.

Leavitt knew to apply this hierarchy of visual concepts because of her training in astronomy. But over the last decade, AI experts have discovered that they can teach computers to extract such conceptual hierarchies directly from raw images, and that this approach is actually far more effective than programming those same computers with domain-specific knowledge, whether of astronomy or cucumbers.

This whole approach is called "deep learning," and until recently it was an academic curiosity. Not anymore; at certain image-identification tasks, deep neural networks now outperform people. AI experts use a data set called the ImageNet Visual Recognition Challenge to benchmark their models. ImageNet is an online database with millions of photos across a thousand different categories, like "sailboat" and "Alaskan Malamute," and the goal is to train a model to automatically identify the images. Humans have an average error rate of about 5% on this task, while in 2011, the best AI models had a 25% error rate. By 2014, however, Google's Inception model had brought the world-record error rate for a machine down to 6.7%. Inception was a 22-layer deep neural network, with each of its layers representing a higher level of visual abstraction, from concepts like "circle" and "edge" at the bottom to concepts like "sailboat" and "Malamute" at the top—all learned organically from the data. And by 2016, follow-up models had achieved a sub-3% error rate, *better than the average human.*

The Potential Benefits . . .

Deep learning represents nothing less than a revolution in the visual competence of machines—and the core ideas and technology are dif-

‡ This is just for images. There are deep neural-network architectures for all kinds of inputs, from videos to text, and they are all structured differently.

fusing everywhere. The limits here were once the availability of data, the speed of our computers, and the richness of our models. Today, there seems to be only one limit, and that's the imagination of people:

- A Swedish beekeeper, Björn Lagerman, is trying to save honeybees with a deep-learning model, trained on 40,000 images of bee colonies, that can alert beekeepers to the presence of Varroa mites, the most dangerous enemy of the Western honeybee.[11]
- Mark Johnson and his startup, Descartes Labs, have trained deep neural networks on four petabytes of satellite imagery, together with crop reports from the U.S. Department of Agriculture, to predict corn yields. These predictions are crucial for the countless merchants along the agricultural supply chain, from the owners of grain elevators to the producers of ethanol. Since 2014, Descartes has consistently outpredicted the USDA.[12]
- Electricity providers are now training models that use weather data to predict grid-level demand for electricity. In England, they're being combined with satellite imaging data to predict the power output from renewable sources, like solar and wind. The National Grid estimates that this kind of deep-learning model could ultimately shave 10% from England's power bill, just by balancing supply with demand more efficiently.[13]

Then there's the recent study published by the Geena Davis Institute on Gender in Media. Researchers at the institute started to collect data back in 2007 on how men and women were portrayed differently in films. They initially did the data analysis by hand, watching thousands of hours of film and looking for patterns scene by scene. But they recently teamed up with Google to automate this task, using deep neural networks for image classification. They used an updated version of the Inception model to analyze the 100 highest-grossing Hollywood films across several years. This model automatically classified the gender of each person onscreen, as well as who was speaking at any given moment.

The results were striking. There's only one genre in which women get more screen time than men: horror films, where they're usually the victims. Across every other genre, women are underrepresented. On average, they receive 36% of the screen time and 35% of the speaking time—and only 27% of the speaking time in films nominated for an Academy Award. These results illustrate how AI can help to inform discussions about gender stereotypes and unconscious bias.[14]

. . . And the Threat to Privacy

While we've highlighted the many potential benefits of these new pattern-recognition algorithms, it's equally important to acknowledge the concerns about privacy that they raise. The same deep-learning technology used to identify gender bias in Hollywood films can also be used by police, for example, to monitor footage from CCTV cameras in public places. With enough cameras and enough training data, there's no technological reason why an AI system couldn't be programmed to track a single person step by step around a large city. To be sure, the police have always sought to surveil people suspected of crimes, whether by steaming open their letters, tapping their phones, or monitoring their cell-phone metadata. What's different in the age of AI is the theoretical potential for them to monitor *everyone*, all at once—something that, all legal constraints aside, is logistically impossible with human intelligence alone. And the potential for abuse doesn't stop with the police. A private company with access to all that footage could use it to build an extraordinarily fine-grained database of what we look at and for how long, or a government official could use it to retrospectively search for embarrassing details that could be exploited to threaten or intimidate someone—like a journalist or a political opponent, for example.

If you read a lot about AI, you'll encounter two overlapping narratives of the surveillance issue. The less extreme version goes something like this. AI face-recognition technology is immensely powerful, and it needs to be regulated carefully, in the same way that we regulate other technologies ripe with potential for abuse. We endorse this view wholeheartedly. There is a vast discrepancy between our AI technology and our laws, and society must address this problem yesterday, not tomorrow. We need smart rules, crafted by people who understand what they're governing—both the benefits and the potential threats.

The more extreme narrative, on the other hand, holds that there's something uniquely and unavoidably autocratic about the surveillance capabilities of modern AI. We're admittedly not experts in the sociology of technology, but we've yet to see good evidence for this claim. The Gestapo certainly didn't need AI to perfect the art of spying—and for that matter neither did the FBI in the 1950s, when it snooped on civil rights organizations. Moreover, if we look globally today, there is no obvious correlation between a country's engagement with digital technology and its respect for privacy and basic human rights. In Scandinavia, for example, you can find both the world's most stringent digital privacy laws *and* the world's most advanced digital economies. (Good luck paying in cash for a coffee in Stockholm.) These facts complicate any effort to draw a simplistic association between AI technology and tyranny.

So bring on the lawyers and the policy-makers. Fears about privacy are well founded, but we are optimistic that they are also solvable.

Postscript

We'll close with one final story about stereotypes—one especially relevant to a world in which only 17% of computer-science majors at American universities are women, a fraction that's been declining for decades.

You'll recall that Edwin Hubble used Henrietta Leavitt's prediction rule for pulsating stars—the standard candles of the universe—to prove conclusively that the Milky Way wasn't the only galaxy out there. In doing so, he settled a question that astronomers had debated for centuries. When he announced his discovery to the world, Hubble became an instant celebrity. Scientists and journalists clamored for his attention. He would go on to win medals and prizes, to walk among movie stars and heads of state, to have Einstein call at his home, and to have a great telescope that orbits Earth named in his honor.

None of these plaudits went to Henrietta Leavitt. She died of cancer in 1921, four years before Hubble announced his discovery. Professional astronomers, all of them men, certainly knew about her breakthrough equation that showed them how to use pulsating stars to measure the size of the universe. As a group, however, they gave her far less credit than she deserved. For too many of them, Leavitt was "only" a computer:

a woman barred from setting foot in an observatory, and who needed a male sponsor before she could be trusted to publish her paper in a reputable journal. And to the public, she was simply anonymous—just as she remains, more or less, today. We can confidently predict that the 100-year anniversary of Hubble's discovery, when it comes around in 2025, will make the headlines in the world's major newspapers. But the 100-year anniversary of Leavitt's discovery, in 2012, didn't even make the headlines in the world's major astronomy journals.

We owe her much more than this. For if pulsating stars are the candles of the universe, then Henrietta Leavitt was the one who fashioned their candlestick, giving us an equation that we could hold up to the heavens to shine light into the darkness.

3

THE REVEREND AND THE SUBMARINE

Q: What do a bicycle, snow, a kangaroo, and a submarine have in common?

A: They're all important for building a car that can drive itself.

TAKE BICYCLES. BICYCLES are trouble. The sensors on a self-driving car are really good at identifying things like pedestrians and squirrels, which move a lot slower than cars and which look basically the same from every angle. Other cars are a piece of cake; they're giant reflective blobs of metal that light up like a Christmas tree on radar. But bicycles? Bicycles can be fast or slow, big or small, metal or carbon fiber—and depending on your viewing angle, they can be as wide as a car or as narrow as a book. If you don't notice the bicycle itself, how can you tell that a bicyclist isn't just a pedestrian with eccentric posture? And don't even start on all that swerving. So erratic and impulsive. It's enough to give a robot a serious headache.

Snow is trouble, too, and not because of traction—robots are smart enough to get winter tires, and they know their own limits. But snow covers lane lines. Snow obscures stop signs. Snow interferes with the laser beams that the car uses to measure distance to nearby objects. To a robot car, snow means sensory deprivation.

As for kangaroos, while other critters may be unpredictable, at least they stay on the ground. But a kangaroo jumps 30 feet at a time. As it bounds up and down, it seems to get bigger and smaller, bigger and smaller in the camera's field of vision, like a giant rabbit in a kaleidoscope. This is really confusing to a robot. With all those rapid changes in apparent size, how can you tell how far away it is? It's almost like you need a dedicated kangaroo-range-finding laser—maybe lots of them, since kangaroos travel in . . . well, zoologists call them "mobs" for a reason.

Then there's a submarine. We promise we'll get to that in a minute.

But first, we invite you to reflect on something. Isn't it astonishing that we're talking about *mobs of kangaroos* as one of the big technological problems here—as opposed to, say, getting out of the driveway, or not crashing into your neighbors' living room? Ask yourself a simple question. If you had to put a loved one in a taxi, would you rather have them driven by a randomly sampled 16-year-old with a driver's license, or by one of Waymo's cars? (Waymo is the autonomous-car company spun out of Google.) If you have to think about this question, we encourage you to consider a few facts.[1]

56% of American teenagers talk on the phone while driving.

In 2015, 2,715 American teenagers died, and 221,313 went to the emergency room, because of car-crash injuries.

Half of all crashes involving teenage drivers are single-vehicle crashes.

In contrast, Waymo's cars never get distracted. They never drink. They never get tired, and they never text their friends when they should be paying attention to the road. Since 2009, they've driven more than 2 million miles on public roads, and in that time they've caused only one accident: a fender-bender with a city bus in California, while traveling two miles per hour. All told, Waymo's per-mile, at-fault accident rate over nine years is 40 times lower than the rate for drivers ages 16–19, and 10 times lower than for drivers ages 50–59. And that's the *prototype*.

These numbers predict a clear trajectory in the cultural norms of the future: the notion of allowing a 16-year-old to drive a car will seem absurdly irresponsible. When our descendants learn that this used to be normal, they'll react the same way people react today upon learning that their grandparents used to drive home without seat belts after drinking four martinis. As for bicycles, snow, and kangaroos? Those are just engineering problems. They'll be fixed in the near future, perhaps even by the time you read this, and almost surely with the same solution: better data. In AI, data is like water. It's the universal solvent.

In fact, if you hang out with enough data scientists who work on self-driving cars, you'll quickly be confronted with a striking question: Has the last Californian to hold a driver's license already been born?

The Robotics Revolution

Robots have come a long way in a short time.

In the 1950s, the state of the art was Theseus, a life-size autonomous mouse built by Claude Shannon at Bell Labs, and powered by a bank of telephone relays. Theseus the ancient Greek hero entered a labyrinth to slay the Minotaur. Theseus the mouse had more modest ambitions: he entered a 25-square tabletop maze to find a block of cheese. At first he would navigate by trial and error until he found the cheese. After this first triumph, he could find his way back to the cheese from anywhere in the maze, without a mistake.[2]

In the 1960s and '70s, there was the Stanford Cart: a wagon-sized vehicle with four small bicycle wheels, an electric motor, and a single TV camera. The Cart began as a test vehicle to study how engineers might control a moon rover remotely from Earth. But it soon morphed into a platform for research on autonomous navigation for a whole generation of robotics students at the Stanford Artificial Intelligence Lab. By 1979, after years of refinement, the Cart could steer itself across a chair-filled room in five hours, without human intervention—quite an achievement for the time.[3]

And today? Self-driving cars are just the start. Don't forget about autonomous flying taxis, like the ones the government of Dubai has been testing since September of 2017. Or the autonomous iron mine run by Rio Tinto, in the middle of the Australian outback. Or the autonomous shipping terminal at the Port of Qingdao, in China—six enormous berths spanning two kilometers of coastline, 5.2 million shipping containers a year, hundreds of robot trucks and cranes, and nobody at the wheel.[4]

One of the most common questions we hear from students in our data-science classes is "How do these robots work?" We'd love to answer that question in all its glory. Alas, we can't. For one thing, there are so many details that it would take a much longer book, one with lots and lots of equations. Besides, a lot of the details are proprietary. You may have heard, for instance, that Waymo has sued Uber for $1.86 billion over the alleged theft of some of those details—a suit whose outcome, at the time of writing, was unknown.[5]

Details aside, though, let's think about the big picture. Here's an analogy. You can certainly learn the basics of how a plane stays in the air, even if you don't know how to build a Boeing 787. Similarly, you can understand how a self-driving car navigates its environment, even if you can't design such a car yourself. That's exactly the level of understanding you'll achieve by the end of this chapter, by building on what you've already learned about conditional probability.

To get there, we'll start with a simple, almost childlike question, one fundamental to any autonomous robot—whether it walks, drives, or flies; whether it digs up iron or ferries us to the grocery store; whether it's the size of a mouse or the size of a container ship. This question, in fact, is so important that it must be asked and answered dozens or even hundreds of times every second.

That question is: Where am I?

In AI, this is called the SLAM problem, for "simultaneous localization and mapping." The word "simultaneous" is key here. Whether you're a person or a robot, knowing where you are means doing two things at once: (1) constructing a mental map of an unknown environment, and (2) inferring your own unknown location within that environment. This is a chicken-and-egg problem. Your beliefs about the environment depend on your location, but your beliefs about your location depend on the environment. Neither can be known independently of the other, so it seems logically impossible to infer both at once. Imagine, for example, that you're trying to get to Times Square having never been to New York before, and we give you directions by telling you that it's one subway stop north of Penn Station—and then, when you ask where Penn Station is, we tell you that it's one stop south of Times Square. Now you're supposed to go find both places without a map. That's the SLAM problem.

Although it may not seem like it to you, the information you get from your senses about your own location in the world is just about as circular as our directions to Times Square. The cognitive miracle is that you can solve the SLAM problem routinely, with no conscious effort, every time you walk into an unfamiliar room. Neuroscientists don't fully understand how you do it, but they do know that there's a lot of very specialized and phylogenetically ancient brain circuitry involved, especially in the hippocampus. And like many capabilities honed by evolution, this one turns out to be really hard to reverse engineer. In AI, this is often called the "Moravec paradox": the easy things for a five-year-old are the hard things for a machine, and vice versa.*

The current revolution in autonomous robots has become possible only because all the research put into SLAM systems has finally paid off. Robots have gone from dodging chairs to dodging other drivers; from five hours traversing a room to five gigabytes of sensor data per second; from an autonomous mouse that can navigate a 25-square grid to an autonomous car that can navigate millions of miles of public road. SLAM is one of AI's most smashing success stories. So in this chapter, we'd like to address two SLAM-related questions—one obvious, and one a bit more unexpected.

1. How does a robot car know where it is?
2. How can you become a smarter person by thinking a bit more like a robot car?

The answer to both of these questions involves something called *Bayes's rule*. Bayes's rule is how self-driving cars know where they are on the road—but it's so much more than that. Bayes's rule is a profound mathematical insight used every day, in almost every area of science and industry. Moreover, it is a supremely helpful principle for living

---

* Named after Hans Moravec, a pioneer in robotics.

your day-to-day life in a smarter way—like, for instance, if you want to invest more prudently, or make more informed decisions about medical care. Bayes's rule provides the single best example of how training yourself to think a bit more like a machine can help you be a wiser, healthier person.

How Is Finding a Submarine like Finding Yourself on the Road?

We can now, at last, return to our earlier promise. We told you that understanding a submarine would be important for getting a car to drive itself. Now let us show you why.

The connection here is Bayes's rule, which we'll explain by telling you a story about a submarine. Not a self-driving submarine or anything like that—just a run-of-the-mill nuclear-powered attack submarine called the USS *Scorpion*. The *Scorpion* is famous because one day, in 1968, it went missing, somewhere along a stretch of open ocean spanning thousands of miles. This sent the U.S. military into a frenzy; it's a big deal when a nuclear submarine vanishes. Despite the long odds, navy officials threw everything they had into the search. They combed the ocean for months on end, yet they couldn't find the *Scorpion*. Discouraged and hopeless, they were about ready to call it quits.

But one man was too stubborn to give up. His name was John Craven, and he was stubborn because he was convinced that he had probability on his side—and the remarkable thing is that he was right. John Craven and his search team used Bayes's rule to answer the question "Where is the *Scorpion* in this big empty ocean?" Once you know how they did it, you'll understand how a self-driving car uses the same math to answer a very similar question: "Where am I on this big open road?"

The Search for the *Scorpion*

In February of 1968, the USS *Scorpion* set sail from Norfolk, Virginia, under the watch of Commander Francis A. Slattery. The *Scorpion* was a Skipjack-class high-speed attack submarine, the fastest in the American fleet. Like other subs of her class, she played a major role in U.S.

military strategy: this was the height of the Cold War, and both the Americans and the Soviets deployed large fleets of these attack subs to locate, track, and—should the unthinkable happen—destroy the other side's missile submarines.

On this deployment, the *Scorpion* sailed east, bound for the Mediterranean, where for three months she participated in training exercises alongside the navy's 6th Fleet. Then, in mid-May, the *Scorpion* was sent back west, past Gibraltar and out into the Atlantic. There she was ordered to observe Soviet naval vessels operating near the Azores—a remote island chain in the middle of the North Atlantic, about 850 miles off the coast of Portugal—and then to continue west, bound for home. The sub was due back in Norfolk at 1:00 P.M. on Monday, May 27, 1968.

The families of the *Scorpion*'s 99 crew members had gathered on the docks that day to welcome their loved ones back home. But as 1:00 P.M. came and went, the sub had not yet surfaced. Minutes stretched into hours; day gave way to night. Still the families waited. But there was no sign of the *Scorpion*.

With growing alarm, the navy ordered a search. By 10:00 P.M. the operation involved 18 ships; by the next morning, 37 ships and 16 long-range patrol aircraft.[6] But the odds of a good outcome were slim. The *Scorpion* had last made contact off the Azores, six days earlier. She could have been anywhere along that 2,670-mile strip of ocean between the Azores and the eastern seaboard. With every passing hour, the chances that the sub could be located, and that rescue gear could be deployed in time, were rapidly fading. At a tense news conference on May 28, President Lyndon Johnson summarized the mood of a nation: "We are all quite distressed. . . . We have nothing that is encouraging to report."[7]

After eight days, the navy was forced to concede the obvious: the 99 men of the crew were declared lost at sea, presumed dead. The navy now turned to the grim task of locating the *Scorpion*'s final resting place—a tiny needle in a vast haystack stretching three-fourths of the way across the North Atlantic. Although hopes for saving the crew had been dashed, the stakes for finding the sub were still high, and not only for the families of those lost: the *Scorpion* had carried two nuclear-tipped torpedoes, each capable of sinking an aircraft carrier with a single hit. These dangerous warheads were now somewhere on the bottom of the sea.

John Craven, Bayesian Search Guru

To lead the search, the Pentagon turned to Dr. John Craven, chief scientist in the navy's Special Projects Office, and the leading guru on finding lost objects in deep water.

Remarkably, Craven had done this kind of thing before. Two years earlier, in 1966, a B-52 bomber had collided in midair with a refueling tanker over the Spanish coast, near the seaside village of Palomares. Both planes crashed, and the B-52's four hydrogen bombs, each of them 50 times more powerful than the Hiroshima explosion, were scattered for miles. Luckily none of the warheads detonated, and three of the bombs were found immediately. But the fourth bomb was missing and was presumed to have fallen into the sea.

Craven and his team had to ponder many unknown variables about the crash. Had the bomb remained in the plane, or had it fallen out? If the bomb had fallen out, had its parachutes deployed? If the parachutes had deployed, had the winds taken the bomb far out to sea? If so, in what direction, and exactly how far? To sort through this thicket of unknowns, Craven turned to his preferred strategy: *Bayesian search*. This methodology had been pioneered during World War II, when the Allies used it to help locate German U-boats. But its origins stretched back much further, all the way to a mathematical principle called Bayes's rule, first worked out in the 1750s.[8]

Bayesian search has four essential steps. First, you should create a map of *prior probabilities* over your search grid. These probabilities are "prior" in that they represent your beliefs before you have any data. They combine two sources of information:

- The presearch opinions of various experts. In the case of the missing H-bomb, some of these experts would be familiar with midair crashes, some with nuclear bombs, some with ocean currents, and so forth.
- The capability of your search instruments. For example, suppose that the most plausible scenario puts the lost bomb at the bottom of a deep ocean trench. Despite its plausibility, you might not actually want to begin your search there: the trench is so dark and re-

mote that even if the bomb were there, you'd be unlikely to find it. To draw on a familiar metaphor, a Bayesian search has you start looking for your lost keys using a precise mathematical combination of two factors: where you think you lost them, and where the streetlight is shining brightest.

You can see an example map of prior probabilities in the top panel of Figure 3.1.

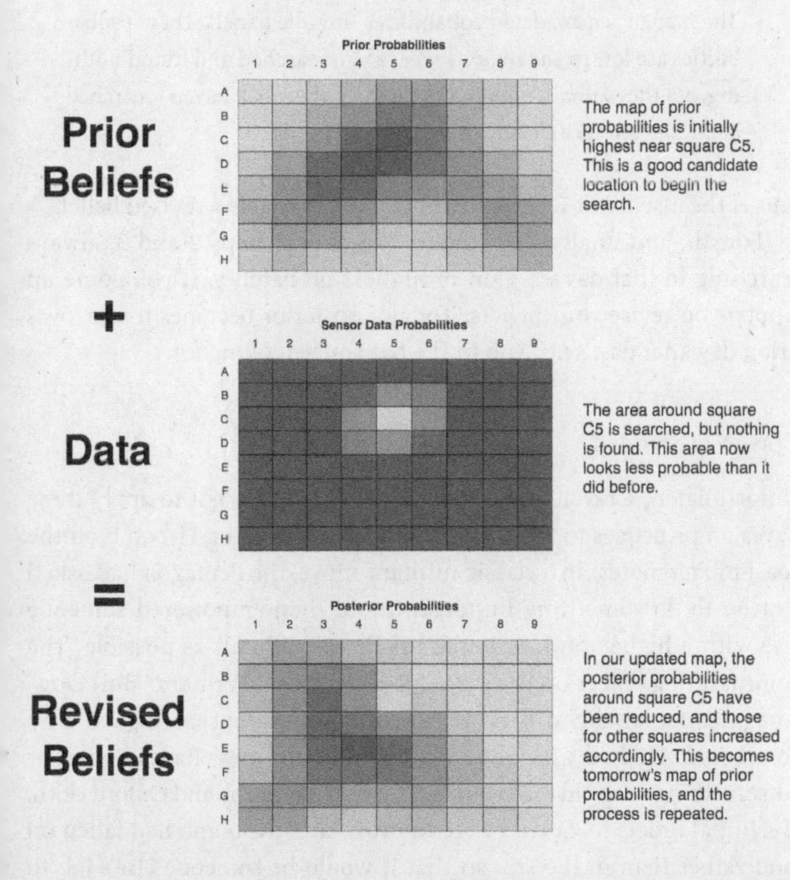

Figure 3.1. In Bayesian search, prior beliefs are combined with data from search sensors to yield a set of revised beliefs.

The second step is to search the location of highest prior probability, which is square C5 in Figure 3.1. If you find what you're looking for, then you're done. If you don't, you move on to the third step: revise your beliefs. Suppose you searched around square C5 but found nothing. Now you reduce the probability around square C5 and bump up the probability in the other regions accordingly. Your prior probabilities have now, in light of the new data, become posterior probabilities. You can visualize this by overlaying two maps on top of each other:

- The original map of prior probabilities (top panel).
- The map of sensor-data probabilities (middle panel). These probabilities are low in the regions where you searched and found nothing, but they remain high in the regions where you haven't searched at all, since you can't rule them out.

This is the essence of Bayes's rule: prior belief + facts = revised belief.

Fourth, and finally, you iterate. You repeat steps 2 and 3, always searching in that day's region of highest probability. If you come up empty, you revise your beliefs. Today's posterior becomes tomorrow's prior, day after day, until you find what you're looking for.

Craven Is Stymied

Unfortunately, Craven and his team never actually got to apply these Bayesian principles to the 1966 search for the missing H-bomb off the coast of Palomares. In a classic military move, the Pentagon had asked Craven to do something important, and then empowered someone else with a higher rank to make his life as difficult as possible. The commanding officer on the scene, Rear Admiral William "Bull Dog" Guest, had a notably different view of the way the search should be conducted. He had little patience for probabilities, Bayes's rule, or 20-something-year-old math PhDs dressed in corduroy and Oxford cloth. His initial orders to Craven were to prove that the bomb had fallen on land rather than in the sea, so that it would be someone else's job to find the damned thing. As a result, the search for the Palomares H-bomb was really two searches. There was Craven's Bayesian "shadow" search, with its slide rules and probability maps, and with updated numbers constantly chattering over the teletype machine as the mathematicians fed remote calculations to a mainframe back in Pennsylvania. But the insights arising from these calculations were ignored in favor of Admiral Guest's "plan of squares," which guided the real search.

Eventually, the Palomares H-bomb was found, after it was discovered that a local fisherman had seen the bomb fall into the water under a parachute and could guide the navy to its point of entry. Thus while the search was a success, the Bayesian part of it had been a failure, for the simple reason that it had never been given a chance. Nonetheless, the Palomares incident taught John Craven some valuable lessons—both about the practicalities of conducting a Bayesian search and about how to get support for that search from the military brass.[9]

Two years later, when he was called upon to find the *Scorpion*, Craven was ready.

The Search for the *Scorpion* Continues

When the *Scorpion* disappeared in May of 1968, Craven and his team of Bayesian-search experts quickly reconvened. At first, their task seemed vastly more daunting than the search for the Palomares bomb. Back then, they'd known to confine the search to a relatively small area in the shallow seas off the coast of southern Spain. Here, the team had to find a submarine two miles underwater, somewhere between Virginia and the Azores, without so much as a single clue.

Luckily, they caught a break. Starting in the early 1960s, the U.S. military had spent $17 billion installing an enormous, highly classified network of underwater microphones throughout the North Atlantic, so that they could track the movements of the Soviet navy. Highly trained technicians at secret listening posts monitored these microphones around the clock. After sniffing around, Craven discovered that one of these listening posts in the Canary Islands had, one day in late May, recorded a very unusual series of 18 underwater sounds. Then he learned that two other listening posts—both of them thousands of miles away, off the coast of Newfoundland—had recorded those very same sounds around the same time. Craven's team compared these three readings and, by triangulation, worked out that the sounds must have emanated from a very deep part of the Atlantic Ocean, about 400 miles southwest of the Azores. This location fell along the *Scorpion*'s expected route home. Moreover, the sounds themselves were highly suggestive: a muffled underwater explosion, then 91 seconds of silence, and then 17 further sonic events in rapid succession that, to Craven, sounded like the implosion of various compartments of a submarine as it sank beneath its hull-crush depth.[10]

This acoustic revelation dramatically narrowed the size of the search area. Still, the team had about 140 square miles of ocean floor to cover, all of it 10,000 feet below the surface, and therefore inaccessible to all but the most advanced submersibles.

The Bayesian search now kicked into high gear. Craven and his team interviewed expert submariners, who came up with nine possible scenarios—a fire on board, a torpedo exploding in its bay, a clandestine Russian attack, and so on—for how the submarine had sunk. They weighed the prior probability of each scenario and ran computer simulations to understand how the submarine's likely movements might have unfolded under each one. They even blew up depth charges at precise locations, to calibrate the original acoustical data from the listening posts in the Canary Islands and Newfoundland.

Finally, they put all this information together to form a single search-effectiveness probability for each cell on their search grid. This map crystallized thousands upon thousands of hours of interviews, calculations, experiments, and careful thinking. It would have looked something like Figure 3.2.

Predictably, Craven encountered both logistical and bureaucratic difficulties in getting the Pentagon to pay attention to his map of probabilities. Summer came and went. By this point, the search for the *Scorpion* had been going for over three months, to no avail.

Eventually his cajoling paid off, and the military brass ordered that Craven's map be used to guide the search. So starting in October, when commanders leading the search aboard the USS *Mizar* finally got hold of the map, the operation became truly Bayesian. Day by day, the team rigorously searched the region of highest probability, crunched the

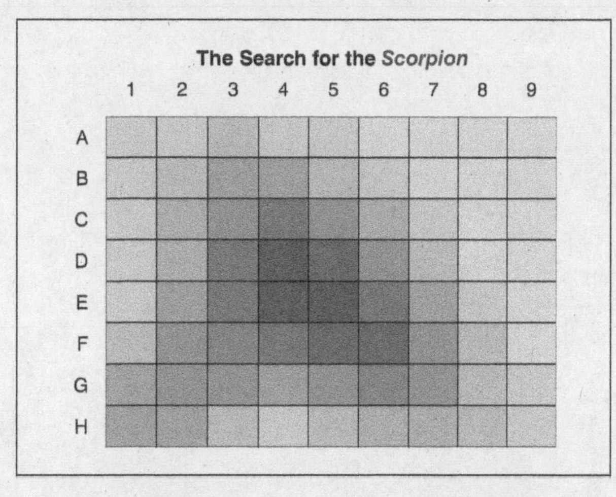

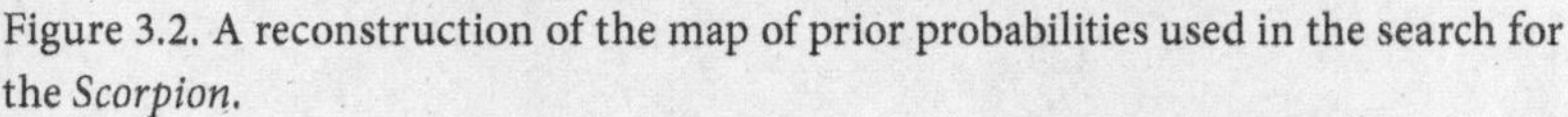
Figure 3.2. A reconstruction of the map of prior probabilities used in the search for the *Scorpion*.

numbers, and updated the map for tomorrow. And day by day, those numbers were slowly homing in on rectangle F6.

On October 28, Bayes finally paid off.

The *Mizar* was in the midst of its fifth cruise and its seventy-fourth individual run over the ocean floor. All of a sudden the ship's magnetometer spiked, suggesting an anomaly on the sea floor. Cameras were hurriedly deployed to investigate—and sure enough, there she was. Partially buried in the sand, 400 miles from landfall, and two miles below the surface of the sea, the USS *Scorpion* had been found at last.[11]

To this day, nobody knows for sure what happened to the *Scorpion*—or if they do, they're not talking. The navy's official version of events cites the accidental explosion of a torpedo or the malfunctioning of a garbage-disposal unit as the two most likely possible causes. Many other explanations have been proposed over the years—and as with any famous mystery, conspiracy theories abound.[12]

But there was at least one definitive conclusion to come out of the incident: Bayesian search worked brilliantly. As it turned out, the sub's final resting place lay a mere 260 yards away from rectangle E5, the initial region of highest promise on Craven's map of prior probabilities.

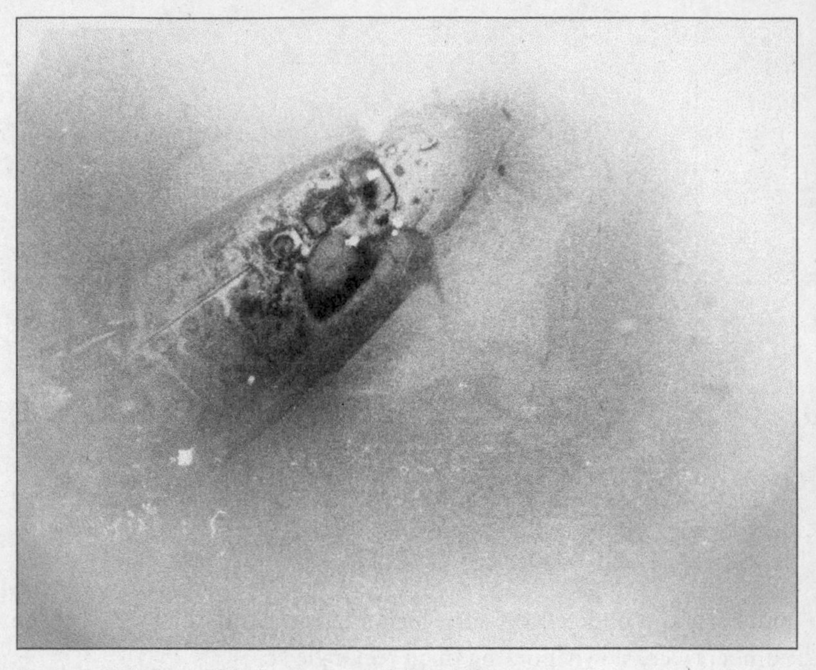

Figure 3.3. A photo of the bow section of the USS *Scorpion*, taken in 1968 by the crew of the bathyscape *Trieste II*.

The search team had actually passed over that location on a previous cruise but had missed the telltale signs due to a broken sonar.[13]

Ponder that for a moment more. Think of how hard it is to find something you've lost on a 100-foot stretch of beach, or for that matter in your own living room. Yet when a lone submarine had disappeared somewhere in a 2,600-mile stretch of open ocean, a Bayesian search had pinpointed its location to within 260 yards, only three lengths of the submarine itself. It was a remarkable triumph for Craven's team—and for Bayes's rule, the 250-year-old mathematical formula that had served as the search's guiding principle.

Bayes's Rule, from Reverend to Robot

Here's the key mantra we must take away from the story of the *Scorpion*: all probabilities are really *conditional* probabilities. In other words, all probabilities are contingent upon what we know. When our knowledge changes, our probabilities must change, too—and Bayes's rule tells us how to change them.

Bayes's rule was discovered by an obscure English clergyman named Thomas Bayes. Born in 1701 to a Presbyterian family in London, Bayes showed an early talent for mathematics, but he came of age at a time when religious dissenters were barred from universities in England. Denied the chance to study math at Oxford or Cambridge, he ended up studying theology at the University of Edinburgh instead. This must have seemed like a cruel barrier to Bayes, just as it did to so many others of his era. But there was a peculiar side effect of this discrimination. Because of its intolerant religious policies, England was home to a surprising number of amateur mathematical societies formed by talented Presbyterians who, like Bayes, were barred from English universities, and who therefore created their own homegrown intellectual communities instead. In his forties, Bayes became a member of one such society, in a spa town in Kent called Tunbridge Wells, where he'd taken a job as a minister—and where he came up with the rule that now bears his name, sometime during the 1750s.

Surprisingly, his discovery didn't make much of an impact at first. Bayes didn't even publish it during his lifetime; he died in 1761, and his manuscript was read posthumously to the Royal Society in 1763, by his friend Richard Price. There was a brief period around the turn of the nineteenth century when Bayesian ideas flourished, mainly in the hands of the great French mathematician Pierre-Simon Laplace. But upon Laplace's death in 1827, Bayes's rule fell into obscurity and irrelevance for more than a century.

Bayesian Updating and Robot Cars

Today, however, Bayes's rule is back, better than ever, and sitting right behind the steering wheel of every robot car out there.

Bayes's rule is an equation that tells us how to update our beliefs in light of new information, turning *prior* probabilities into *posterior* probabilities. It offers the perfect solution to the robotics problem we discussed earlier: SLAM, or simultaneous localization and mapping.

SLAM is an inherently Bayesian problem. As new sensor data arrives, a robot car must update its "mental map" of the surrounding environment—the lane markers, the intersections, the traffic lights, the stop signs, and all the other vehicles on the road—while simultaneously inferring its own uncertain location *within* that environment. In essence, a robot car "thinks" of itself as a blob of probability, traveling down a Bayesian road.

Before we describe how this works, let's address one obvious question up front: Why not just navigate using GPS technology, like the kind you have on your smartphone? The problem is that even under ideal conditions, civilian-grade GPS systems are accurate only to about 5 meters—and they're much worse in tunnels or near tall buildings, where they might err by 30 or 40 meters. Trying to navigate city traffic using GPS alone would be like trying to perform vascular surgery while wearing oven mitts and a blindfold.

So to supplement the information it gets from its GPS receiver, a robot car must turn to a bevy of other sensors. Some of these are plain old video cameras, while others are just like the safety features on most new cars today—for example, bumper-mounted radar, like the kind that beeps at you whenever you're in danger of backing up into something.

A robot car's coolest and most useful sensor is called LIDAR, a portmanteau of "light" and "radar" that stands for "light detection and ranging." Imagine being blindfolded and told to make your way across an unfamiliar room with the help of nothing but a cane. You would probably do this by touch: that is, by using the cane to poke and prod around you, informally measuring the distance to things in your immediate vicinity. If you did this enough times, in all different directions, then you could build up a good mental map of your surroundings.

A LIDAR array works on the same principle: it shoots a laser beam and measures distance by timing how long it takes the light to bounce back. A typical LIDAR array might have 64 individual lasers, each sending out hundreds of thousands of pulses per second. Each laser beam provides detailed information about a very specific direction. So to allow the car to see in all directions, the LIDAR is mounted on the roof, in a rotating assembly that spins roughly 300 times per minute, just like

Figure 3.4. A LIDAR image of a highway, courtesy of Oregon State University.

a faster version of the rotating beam on the radar screens from *Top Gun*. The lasers will therefore point in any one direction about five times every second, giving the car *discrete* rather than *continuous* positional updates. In other words, the car sees a world lit not by steady sunshine but by a strobe light—by short flashes of data from its LIDAR and other sensors, each one giving the car a new view of its surroundings, like in Figure 3.4.

Every time the car receives a new burst of data, it uses Bayes's rule to update its "beliefs" about its location. We can visualize this Bayesian updating process using a map where the road is broken up into little grid cells, each with its own probability. Suppose that you've pulled out of the driveway in your autonomous car, and that you're now 60 seconds into your journey, traveling at about 30 miles per hour. Based on the data up to now, the car has a set of beliefs about its position. This is shown as a map of probabilities in the upper-left panel of Figure 3.5. Now let's check in with the car one-fifth of a second later, after one sweep of the LIDAR array, 60.2 seconds into your journey. How have those beliefs changed?

There are three steps in the car's reasoning. The first is what

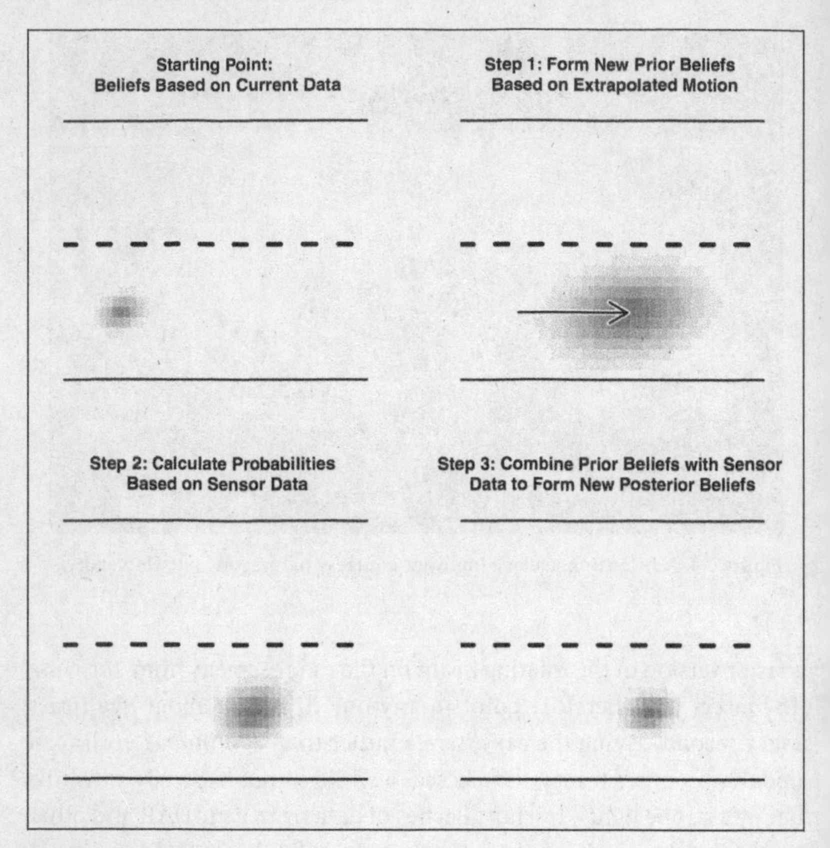

Figure 3.5. How an autonomous car uses Bayes's rule to update its beliefs about its location.

navigation experts call "dead reckoning," or as we like to call it, "introspection and extrapolation." *Introspection* means collecting "internal state" information, like speed, wheel angle, and acceleration; *extrapolation* means using this information, together with the laws of physics, to forecast the probable movement of the car over the next fraction of a second. The result is a map of prior probabilities for the car's location 60.2 seconds into the journey, shown in the upper-right panel of the figure above. These probabilities are "prior" because they don't yet incorporate updated sensor data; we're in the instant just before the next flash of the strobe.

You'll notice two things here: the blob of probability has moved down the road a bit, and it's been "smeared out" to cover a larger area. This smearing represents the additional uncertainty introduced by extrapolation. For example, if you're traveling 30 miles per hour, you'd expect to cover about nine feet in 0.2 seconds. But you might actually cover a bit more or less as a result of unanticipated steering, braking, or acceleration.

The car's second step is to gather data from its external sensors, like its cameras and LIDAR. These provide a reality check on the car's position, helping to correct the errors introduced by extrapolation. This information is shown in the bottom-left panel of Figure 3.5. You can think of this map as a set of "sensor-only" probabilities—that is, what would the car think about its position based on external sensors alone, in the absence of any prior information?

The car *does* have prior information, though, and so the third and final step is synthesis. Using Bayes's rule, the prior probabilities based on extrapolation (from step 1) are combined with the sensor data (from step 2). In the bottom-right panel, you can see this new map of posterior probabilities, which provides a revised answer to the fundamental question: "Where am I?" Crucially, the blob of posterior probabilities is less smeared out than either the prior or the sensor-data probabilities in isolation. Two sources of information typically mean less uncertainty than you'd have from either source alone.

We've left out a lot of details here. Here's the biggest one: in the example above, we pretended that the road was a fixed reference frame, and that the only unknown variable was the car's location within that frame. That's the *L* in SLAM, for "localization." But don't forget the *M*, for "mapping." In reality, the road itself is unknown, and all its features are subjected to the same Bayesian treatment. Road boundaries, lane lines, pedestrians, other cars, even kangaroos—all are represented as blobs of probability whose locations are constantly being updated with every flash of data from the sensors.

How Bayes's Rule Can Make You Smarter

Viewed through the lens of Bayes's rule, finding a lost submarine and finding yourself on the road turn out to be very similar problems. But

Bayes's rule is far bigger than that. In fact, in terms of its applicability to everyday life, it's one of the most useful equations ever discovered—a perfect mathematical dose of antidogmatism that tells us when to be skeptical and when to be open-minded. Think of all the new information you encounter every day. Bayes's rule answers a very important question: When should that information change your mind, and by how much?

You might never in your life actually sit down with pencil and paper to work through the math of Bayes's rule, and that's totally fine. The point is that, even if you don't, learning to think about the world a bit like a Bayesian car—in terms of priors, data, and how to combine them—can help you be a wiser person. Here are two key examples.

Bayes's Rule in Medical Diagnostics

We'll start with an example with some clear numbers attached—and where even highly trained experts tend to get the answer wrong, because they have failed to apply Bayes's rule.

Imagine that you are a doctor, and that a 40-year-old woman named Alice comes into your office for a routine screening mammogram. Unfortunately, her mammogram result comes back positive, indicating that she may have breast cancer. But you know from your medical training that no test is perfect, and that Alice may have gotten a false-positive result. What should you tell her about the probability that she has cancer, given her positive mammogram? Here are some facts to help you judge.

- The prevalence of breast cancer among people like Alice is 1%. That is, for every 1,000 40-year-old women who have a routine mammogram, about 10 of them have breast cancer.
- The test has an 80% detection rate: if we give it to 10 women with cancer, it will detect about 8 of those cases, on average.
- The test has a 10% false-positive rate: if we give it to 100 women without breast cancer, it will wrongly flag about 10 of them, on average.

In light of these numbers, what is the posterior probability *P*(cancer | positive mammogram)?

The answer, according to Bayes's rule, is quite small: only 7.4%. This number may surprise you. If so, you're not alone: a shockingly high number of doctors guess something much larger. In one famous study, 100 doctors were given the same information you've just been given, and 95 of them estimated that *P*(cancer | positive mammogram) was somewhere between 70% and 80%.[14] They didn't just get the answer wrong; they were off by a factor of 10.

This example raises two questions. First, why is the posterior probability *P*(cancer | positive mammogram) only 7.4%, despite the fact that the mammogram is 80% accurate? Second, how could so many doctors get the answer so badly wrong?

The answer to the first question is this: most women who test positive on a mammogram are healthy, because the vast majority of women who receive mammograms in the first place are healthy. Put simply, *cancer has a low prior probability.* We can visualize this using a *waterfall diagram,* which is like an "everyday life" version of the probability map that a self-driving car uses to navigate the road. In Figure 3.6, we follow a hypothetical cohort of 1,000 women, each 40 years old, as they receive routine screening mammograms. The left branch shows 10 women (1% of 1,000) who actually have breast cancer. Since the test is 80% accurate, we expect that of these 10 cases, 2 will be missed and 8 will be caught. Meanwhile, the right branch shows 990 patients who are cancer-free. Since the test has a 10% false-positive rate, we expect that about 890 will be cleared and 100 will be wrongly flagged, rounding off slightly.[15]

So at the bottom of the waterfall, we end up with 1,000 cases, broken down as follows.

- 108 positive mammograms. Of these, 8 are true positives, or cases of cancer that were detected. The remaining 100 are false positives, or healthy women wrongly flagged by the test.
- 892 negative mammograms. Of these, 2 are false negatives, or missed cancer cases. The other 890 are true negatives, or women correctly given a clean bill of health.

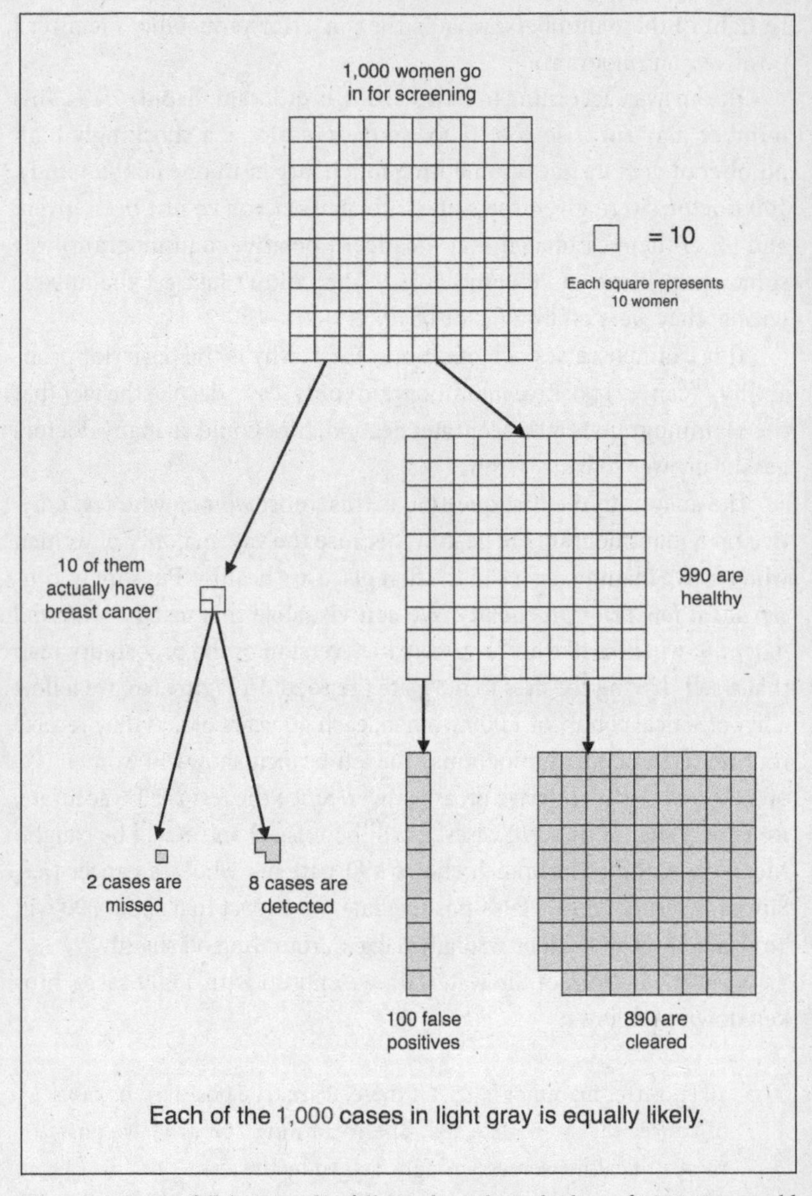

Figure 3.6. A waterfall diagram that follows a hypothetical cohort of 1,000 40-year-old women as they receive routine screening mammograms.

Each of these 1,000 cases is equally likely, so we color in their squares with the same light gray. The fact that cancer is relatively unlikely is reflected not by shading but by sheer numbers: only 10 of these 1,000 squares correspond to cases of cancer.

Now let's use this diagram to think about the situation for your hypothetical patient, Alice. When she first walks into the doctor's office, you know that Alice will be like one of the 1,000 women at the bottom of the waterfall diagram. You just don't know which one. When her mammogram comes back positive, now you know that Alice must be like one of the 108 women with the same result. So let's revisit the diagram and color those 108 cases in a darker gray, while "zeroing out" the other 892 cases by coloring them in white, in Figure 3.6.

Of these 108 positive mammograms, 8 are real cancer cases, while 100 are false positives. Therefore the posterior probability that Alice has cancer, $P(\text{cancer} \mid \text{positive mammogram})$, is about $8/108 \approx 7.4\%$.

And that's Bayes's rule. The prior probability of cancer was 1%. After you see the data, the posterior probability of cancer becomes 7.4%—a lot higher than the prior, but still a far cry from the 70–80% figure estimated by most doctors. (If you'd like to see this worked out using an actual equation, see the sidebar at the end of the chapter.)

Now let's turn to the second question we posed earlier. When asked

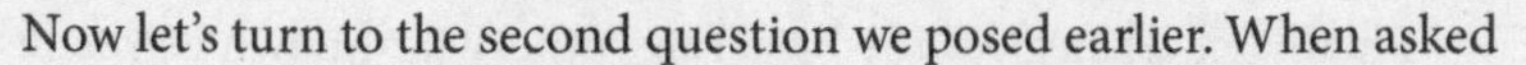

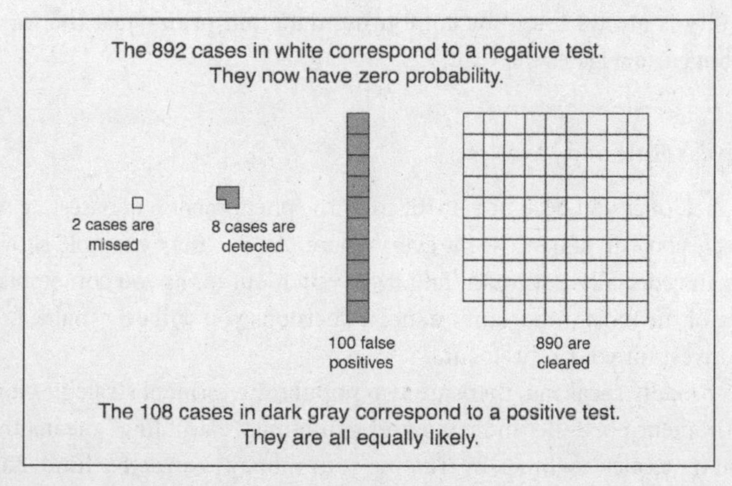

Figure 3.7.

to estimate the posterior probability *P*(cancer | positive mammogram), why did so many doctors come up with a figure 10 times too high? Basically, the doctors were ignoring the prior probability, a fallacy called "base-rate neglect." The doctors' estimates of 70–80% weren't accounting for the low rate of cancer in the population (1%), which implies that most positive tests will be false positives. Instead, the doctors were focusing on just one number: the fact that the test is "80% accurate," meaning that it detects 80% of actual cancer cases. They were giving too much credence to the data, and not enough credence to the prior.

There are three morals to this story. First, never ask your doctor, "How accurate is this test?" At best, you'll get the right answer to the wrong question. Instead ask, "What's the posterior probability that I have the disease?" (Be prepared for a scowl, though, since your doctor may not know what this is.)

Second, although Bayes's rule is conventionally expressed as an equation, you rarely need this equation to calculate a posterior probability. Instead, you can just make a waterfall diagram like the one on the previous page, following a hypothetical cohort of subjects through some data-collection process. You'll get to enjoy the Bayesian omelette without cracking the mathematical egg.

Finally, never neglect the base rate, otherwise known as the prior, when interpreting data. Bayes's rule says that the right posterior probability is always found by combining data *and* prior—just the way a robot car navigates the road.

Bayes's Rule and Investing

In fact, once you become sensitized to the phenomenon of base-rate neglect, you will start to see it everywhere. As our next example shows, it's an especially important fallacy to watch out for as you contemplate one of the most important financial decisions you will ever make: how to invest for your retirement.

Broadly speaking, there are two popular investment strategies for a retirement portfolio: indexing and gambling. "Gambling" means that you try to pick a winner, by trusting your money to an active fund manager who attempts to outperform the market. "Indexing" means that you give up on trying to *beat* the market and instead just *buy* the market, in the form of a broad-based index of stocks like the S&P 500.

Proponents of the gambling strategy argue that it's entirely possible to beat the market over the long run. Their best argument for that claim is exactly two words long: Warren Buffett. Buffett, also known as the "Oracle of Omaha," stands alone in the history of investing. His performance numbers are staggering: from 1964 to 2014, an investor in Buffett's holding company, Berkshire Hathaway, would have turned $10,000 into $182 million. Equally remarkable is Buffett's consistency: his stock picks have outperformed the S&P 500 over nearly every contiguous five-year period since the mid-1960s. And while Buffett is the most famous Wall Street success story, there have also been others—a handful of true market wizards, from Joel Greenblatt to Peter Lynch, whose track records are way too impressive to be attributable to blind luck. The investors who have identified and trusted these extraordinary fund managers have been richly rewarded.

Yet stacked against the example of these rare geniuses, we find a harsh numerical reality: most fund managers are nothing at all like Warren Buffett. Their performance numbers were especially damning over the 10-year period from 2007 to 2016, a decade that included a historic crash followed by a roaring bull market. These were ideal conditions for smart stock pickers. Yet according to Standard and Poor's, 86% of actively managed stock funds underperformed their benchmark indices over this period. In Europe, the picture was worse: 98.9% of domestic stock funds, 97% of emerging-market funds, and 97.8% of global stock funds underperformed, net of fees. The active fund managers in the Netherlands took the global prize: 100% of them failed to beat their benchmarks.[16]

The upshot is that there's real stock-picking talent out there, but it's hard to find. So how should these facts affect your investment strategy? Should you settle for an index fund? Or should you gamble on greatness, in the hopes that you can find one of those rare fund managers who truly can beat the market?

If you decide to gamble, then you should be honest with yourself about your goal: to conduct your own Bayesian search for the next Warren Buffett. The possible "search locations" are all the different fund managers competing for your capital, and your "search data" is the information on each manager's track record. What are the chances that your search can locate one of the rare exceptions, in a vast ocean of fund managers who cannot consistently beat the market?

Alas, Bayes's rule gives a pretty clear answer: the chances are terrible.

To show why, we'll appeal to a metaphor that makes the question easy to approach using probability: most mutual fund managers are just flipping coins. In some years they flip heads, and they beat the market. In other years they flip tails, and the market beats them. (Of course, regardless of whether they flip heads or tails, they still charge you fees.) But a rare investor like Warren Buffett is, metaphorically, flipping a coin with heads on both sides. He beats the market year after year, without exception.

Under this metaphor, if we were to compare the 10-year performance of Warren Buffett with that of five regular stock pickers, we might see something like the table below.

	Year 1	Year 2	Year 3	Year 4	Year 5	Year 6	Year 7	Year 8	Year 9	Year 10	Total H's
Jane Doe	H	H	H	T	H	T	T	H	T	T	5
John Bull	T	T	H	H	H	H	T	T	T	H	5
Jean Dupont	H	H	T	H	H	H	H	T	T	H	7
Jan Jansen	T	T	H	H	H	T	T	T	H	T	4
Max Mustermann	T	T	H	H	T	T	T	T	T	H	3
Warren Buffett	**H**	**H**	**H**	**H**	**H**	**H**	**H**	**H**	**H**	**H**	**10**

For our five average investors, their performance is just random. But for Buffett, his performance is driven by his superior stock-picking ability—that special two-headed coin he keeps squirreled away in a safe in Omaha, Nebraska. As you can see from the table, his excellent performance clearly stands out from the crowd.

But here's the problem: on Wall Street, you have to stand out from a much bigger crowd. There aren't merely five mediocre fund managers out there, flipping coins and charging fees. There are thousands upon thousands of them—and the chances are pretty good that at least a few of them will enjoy long winning streaks, just by chance.

That's where Bayes's rule comes in. Imagine a jar containing 1,024 normal coins. Into this jar, a friend places a single two-headed coin. Your friend then gives the jar a good shake, and you draw a single coin at random. You want to know whether your coin has two heads, but it's against the rules to simply look at both sides: you wouldn't be able to do that back in the real world, since every fund manager out there will have some great marketing pitch that makes them sound like a two-headed coin. So you're forced to conduct a *statistical* test for two-headedness, by flipping the coin 10 times.

Now suppose the coin comes up heads on all 10 flips. In light of the evidence, are you holding the two-headed coin or one of the 1,024 ordinary coins? To answer this question with Bayes's rule, let's consider the following facts:

- There are 1,025 coins in the jar: 1,024 are normal, and 1 has two heads.
- The single two-headed coin is guaranteed to come up heads 10 times in a row.
- Any randomly chosen normal coin has probability 1/1024 of coming up heads 10 times in a row. (This is calculated by multiplying ½ by itself 10 times.) Therefore, of the 1,024 normal coins in the jar, we'd expect 1 to come up heads 10 times in a row.

We can tabulate all this information as follows:

	At least 1 tail	10 heads in a row
Normal coins	1,023 (true negatives)	1 (false positives)
Two-headed coin	0 (false negatives)	1 (true positives)

This matrix tells us that of the 1,025 coins in the jar, we'd expect two of them to come up heads 10 times in a row. Only one of them is actually the two-headed coin. There's just a 50% chance that you're holding it, even after 10 flips.

Now let's compare this coins-in-a-jar scenario with the kind of marketing language you might hear from a stock picker at an actively managed mutual fund with an above-average track record:

> Look at my past performance. I've been running my fund for 10 years, and I've beaten the market every single year. If I were just one of those average stock pickers at an inferior fund, this would be very unlikely: less than one chance in a thousand.

The math of this scenario is exactly the same as the one involving the big jar of coins. Metaphorically, the fund manager is claiming to be a two-headed coin, on the basis of flipping 10 heads in a row: beating the market every year for 10 years. But from your perspective, things are not so clear. You should recognize that the clever marketing pitch is implicitly conflating two different probabilities: *P*(10-year winning streak | good stock picker) with *P*(good stock picker | 10-year winning streak). But remember our key lesson from the story of Abraham Wald: conditional probabilities aren't symmetric like that.

So is the fund manager lucky or good? Let's run through the Bayesian calculation under two different prior assumptions. First, suppose you believe that 1% of all stock pickers are true market beaters, and that the other 99% are just flipping coins. Under these assumptions, let's imagine following a cohort of 10,000 stock pickers over 10 years.

- All 100 excellent stock pickers (1% of 10,000) will beat the market every year.
- A mediocre stock picker has about a 1-in-1,000 chance of beating the market 10 times in a row.† Since there are 9,900 mediocre stock pickers, we'd expect about 10 of them to beat the market all 10 years, just by chance.

So that's 110 market beaters, of whom 100 were good and 10 were lucky. Therefore, the posterior probability *P*(market beater | 10-year winning streak) is 100/110, or about 91%.

What if, however, you believed that excellence were much rarer, like

† We've rounded off all the numbers here.

P(true market beater) = 1/10,000? Under this prior, the posterior probability ends up being much lower:

- Now there's only 1 excellent stock picker who beats the market every year.
- Of the 9,999 mediocre stock pickers, we'd again expect about 10 of them to beat the market 10 times in a row, just by chance.

So we have *P*(market beater | 10-year winning streak) = 1/11, or about 9%.[17]

Bayes's rule implies that the right reaction to an investor's track record depends strongly on the prior probability: whether excellent fund managers are common or rare. Yet all available evidence suggests that true market beaters are exceedingly rare. Recall all those ugly statistics on how few funds exceed their benchmarks in even a *single* year, much less 10 years in a row.

For the everyday investor, this has one important consequence. There might indeed be great stock pickers out there. But Bayes's rule implies that, without a *very* long track record, these geniuses cannot be distinguished reliably from the much larger group of mediocrities who are just getting lucky. Even the genius of Warren Buffett became apparent only over a period of decades. Thus when it comes to the question of searching for a talented fund manager, the lesson of Bayes's rule is: don't bother trying. It's even harder than finding a lost submarine in 2,600 miles of open ocean. You'd almost surely be better off investing in a broad index of stocks and bonds, rather than trying to pick winners.

Nevertheless, hope springs eternal. So in case you find your optimism undiminished by the hard reality of Bayes's rule, we'll leave you with this thought. If you're hoping to find the next Warren Buffett, you will only have a marketing pitch to go by—early in a stock picker's career, the performance data is almost useless. So tread carefully, or you'll end up backing the manager with the silver tongue, rather than the golden edge.

Postscript

We first met Bayes's rule as a principle for finding a lost submarine, and today Bayesian search is a small industry, with entire companies that consult on search-and-rescue operations.[18] For example, you might recall the tragedy of Air France Flight 447, which crashed in the Atlantic Ocean on its way from Rio de Janeiro to Paris, in June of 2009. By late 2011, the search for the wreckage had been going on for two fruitless years. Then a Bayesian search firm was hired, a map of probabilities was drawn up, and the plane was found within one week of undersea search.[19]

Moreover, the main idea of Bayes's rule—updating your prior knowledge in light of new evidence—applies everywhere, not least behind the wheel of a self-driving car. Biologists use it to help understand the role of our genes in explaining cancer. Astronomers use it to find planets orbiting other stars on the outer fringes of our galaxy. It's been used to detect doping at the Olympics, to filter spam from your in-box, and to help quadriplegics control robot arms directly with their minds, just like Luke Skywalker.[20] And as you've seen, it's essential for navigating the treacherous landscapes of health care and finance.

So Bayes's rule is much more than just a principle for finding what has been lost. Yes, it helped find the *Scorpion*, and it helps self-driving cars find themselves on the road. But it can also help you find wisdom in confronting the flood of information you face every day.

SIDEBAR: BAYES'S RULE AS AN EQUATION

In most everyday situations, you don't need to know the actual equation of Bayes's rule in order to apply the underlying logic. Maps and waterfall diagrams, like the kind in this chapter, can get you very far, with very little math at all. But if you're interested in a career in data science, or if you just like to understand the details, it's nice to see that equation. So here's Bayes's rule the way it's taught in a college course on AI or statistics.

(continued on next page)

We'll use the letter H to represent a hypothesis that might be true or false, and the letter D to represent some relevant data. Bayes's rule tells us how to use the data to turn the prior probability of the hypothesis, $P(H)$, into a posterior probability, $P(H|D)$:

$$P(H \mid D) = \frac{P(H) \cdot P(D \mid H)}{P(D)}$$

In our medical-testing example, H is the hypothesis that a given patient has breast cancer, and D is the data that her mammogram has come back positive. We know that 1% of patients have breast cancer: $P(H) = 0.01$. Similarly, we know that the test is 80% accurate at detecting breast cancer, if it's present: $P(D|H) = 0.8$. The last thing we need is $P(D)$, the overall probability of a positive test. From our waterfall diagram, we know that out of 1,000 tests, about 108 of them will come back positive: 8 true positives and 100 false positives. Thus $P(D) \approx 108/1000 = 0.108$.

And that's all we need. Let's plug these three numbers into Bayes's rule to calculate the posterior probability of cancer, given the positive test:

$$P(H \mid D) = \frac{0.01 \cdot 0.8}{0.108} = 0.074$$

This is the same probability (7.4%) we calculated from the waterfall diagram.

4

AMAZING GRACE

From Babel to bits: how machines learned to speak our language.

FOR AS LONG as humans have been trying to make machines understand language, those machines have been making silly mistakes. You've surely battled the autocorrect feature on your phone. Or maybe you've traveled abroad and seen how internet translation services can lead people astray, whether at the zoo ("Do not feed the animals; give all food to the guard on duty") or the dry cleaners ("Drop your trousers here"). And the running joke among AI experts is that if Stanley Kubrick had made his film *2001: A Space Odyssey* today, then the conversation between Dave and the malevolent supercomputer HAL-9000 might have gone like this:

```
                    Dave
Open the pod bay doors, HAL.

                    HAL
I've searched the web and found some results
for iPods, Dave. Would you like to see them?
```

Machines make subtler errors, too. IBM's Watson supercomputer was once put through a rhyming test ahead of his big match against human contestants on *Jeopardy!* One test clue was "a boxing term for a hit below the belt." The right rhyming response was "low blow"—but Watson answered "wang bang," a phrase that did not appear in his database and that he must have conceived on his own.

So go ahead, pile on the insults. Still, we encourage you to keep in mind two facts. First, people also make mistakes with language. People spread barbarisms like "for all intensive purposes" or "at his beckon call." People misinterpret song lyrics, like Billy Joel's ("We didn't start the fire, it was always burning, said the worst attorney") or Madonna's ("Like a virgin, touched for the thirty-first time"). People make translation errors, too. Back in 2009, for example, Secretary of State Hillary Clinton made an elaborate show of presenting the Russian foreign minister with a gift: a big red button that was meant to say "Reset" in both English and Russian, to symbolize the Obama administration's policy of "pressing the reset button" on relations with Russia. The policy didn't work out so well, though—and neither did the gift, which didn't say "Reset" in Russian after all, but "Overcharge."

The second thing to keep in mind is that machines are getting better at language—fast. (You must admit that "wang bang" is a creative piece of boxing commentary.) Experts in AI use the term "natural language processing," or NLP, to describe how we get computers to work with language. Over the last few years, you've been living through a period of tremendous growth in successful NLP systems:

- Digital assistants like Amazon's Echo and Google Home are *far* better than the clunky speech-to-text programs of just a few years ago. They can schedule appointments, make a grocery list, choose a song, or rack up charges on your credit card—all by voice, at a level of transcription accuracy that until recently would have seemed like science fiction.
- The version of Google Translate that went live in 2016 represented a massive improvement over previous efforts at machine translation. The software can now generate respectable translations for

over 100 languages—many of them directly from your smartphone camera, like for a restaurant menu or a sign in a train station. Skype can do something similar during a video chat, in real time.

- Chatbots—software designed to simulate human conversation—are becoming a pervasive feature of the digital world. They're especially popular on Facebook Messenger, where you can ask a bot to book a trip through Kayak or pester a merchant to check the status of a delayed package. Chatbots are even more popular in China, where most startup companies create an official bot on WeChat—user base, 930 million people—even before they create a web page.

Machines today are even learning to write. The Associated Press has started using an algorithm that can write a passable recap of a baseball game from a box score, which it currently uses for faraway college games with no reporter present. The system has even learned to insert sports-writing clichés; it just takes the data one game at a time. Data scientists at Salesforce recently developed a similar program that can accurately summarize long articles to help the company's employees digest news reports more quickly. And as two academics who've suffered the slings and arrows of peer review, we were not at all surprised to learn of an algorithm created by researchers at the University of Trieste—one that wrote fake peer reviews good enough to fool real journal editors.[1]

Then there's the side project of software developer Andy Herd, who trained a neural network with a bunch of scripts from *Friends*, the popular 1990s sitcom, to see what kind of new episodes it might write. Sure, the results are nonsense, but they're remarkably *Friends*-like nonsense. Monica is weirdly aggressive, Chandler whines a lot, and there are even cameos by random film stars from the 1990s:

Van Damme

I'll go in a crap.

Monica

Keep talking!

```
                    Phoebe
Wow lady! You're just gonna come over to him
jumpy . . .

                    Chandler
So, Phoebe likes my pants.

                    Monica
Chicken Bob!

                    Chandler
(In a muffin) (Runs to the girls to cry)
Can I get some presents?[2]
```

Imagine what you might get if there were more than 236 episodes—a long run for a sitcom, but a tiny amount of training data by neural-network standards.

So if you want to understand a future with language-aware AI systems, then the interesting question isn't about the sometimes-laughable mistakes these machines make. Rather, it's how they've learned to listen, speak, and even write so effectively.

A Tale of Two Revolutions

There are really two revolutions to talk about here. There's the Programming Language Revolution, which culminated in the 1950s. Then there's the Natural Language Revolution, which we're living through now. These two revolutions differ in important ways, but there's one big idea that unites them: to get a machine to understand words, you have to represent those words in a language the machine can work with. That means you have to turn words into numbers.

For decades, the only effective way to do this was a top-down approach based on prespecified rules. Think of these rules as a contract describing how two parties, the "machine" and the "human," can use language to interact. Imagine the most detailed legal contract you can think of, written by the highest-paid lawyers out there. Then make it a hundred times more detailed.

- There's a set of rules for the human, called a programming language. (Popular examples are Python, Java, and Perl.) A programming language has math symbols like + and =, along with a limited vocabulary of English words, usually rendered in a fixed-width font to intimidate people: `IF, THEN, WHILE,` etc. The language also has a grammar: rules for combining the words into legal "sentences" that instruct the machine to do something specific.
- Then there's a set of rules for the machine, codified in something called a "compiler."[3] These rules operate behind the scenes and are invisible to the human programmer. They provide the machine with detailed step-by-step instructions for translating every conceivable sentence from the programming language into its own internal "machine language" of bits and vectors.

This contract is interpreted in the most literal way possible. If you write a grammatical sentence in the programming language, then the machine has to do *exactly* what you say. And if you deviate from the grammar in even the tiniest of ways, like by misspelling a word or forgetting a semicolon, then the machine basically just gives you the middle finger, or as we like to write it, 00100.

For decades, these were the only terms under which people and computers could have a successful conversation. As you'll learn in this chapter, they're a huge improvement over the way things were at the start of the computer age, when people were forced talk to computers in their native "binary" language of 0s and 1s. But these terms hardly let us use our full powers of language to get our message across. Of course, we can also get computers to do a few trifling things by pointing, clicking, swiping, etc. But that's just so *crude*. Imagine if you had to communicate with other people only by pointing at stuff, or using menus that dropped down from their eyebrows. Language is much more effective—and since the 1950s, if you wanted to use language to *really* boss a computer around, you were stuck with a programming language.

But not anymore. Since about 2010, the brightest minds in AI have forged a second set of contractual terms—a "New Deal" for human-machine linguistic interaction. This New Deal is bottom-up rather than top-down. We start by throwing out the big book of prespecified grammar rules. Instead, we get to talk to machines in our own natural language: English, Chinese, Korean, anything. And the machines have to interpret what we mean, and answer back in our language of choice, without some lawyer in their ear telling them they can ignore us if we miss a semicolon.

To forge this New Deal, we gave the machines three things, whose significance we'll explain in this chapter.

1. Toys: fast GPUs and lots of memory.
2. Fancy software, in the form of neural networks based on "word vectors," a really cool idea at the intersection of language and math that lets us turn words into numbers, so that we can use them to build prediction rules.
3. Above all, the treasure trove of data that's become available over the last two decades, as humanity's linguistic output has gone overwhelmingly digital.

This last one is the most important. People rely on billions of language facts, most of which they take for granted—like the knowledge that "drop your trousers" and "*drop off* your trousers" are used in very different situations, only one of which is at the dry cleaner's. Knowledge like this is hard to codify in explicit rules, because there's too much of it. Believe it or not, the best way we know to teach it to machines is to give them a giant hard drive full of examples of how people say stuff, and to let the machines sort it out on their own with a statistical model.

This purely data-driven approach to language may seem naïve, and until recently we simply didn't have enough data or fast-enough computers to make it work. Today, though, it works shockingly well. At its tech conference in 2017, for example, Google boldly announced that machines had now reached parity with humans at speech recognition, with a per-word dictation error rate of 4.9%—drastically better than the

20–30% error rates common as recently as 2013. This quantum leap in linguistic performance is a huge reason why machines now seem so smart. One might argue, in fact, that human-level speech recognition is the last decade's single most important breakthrough in AI.

So when was the tipping point, and how did we get there? What are "word vectors," and why are they so useful? Why is data so important here—why can't you just get a machine to follow linguistic rules that we write down explicitly, the same way you teach a third-grader to understand English grammar, or a machine to understand Python?

To answer these questions, we'd like to tell you the story of Grace Hopper. She was nicknamed "Amazing Grace," and not just because she's the only person in this book to have appeared on the David Letterman show.* Hopper earned a PhD in mathematics from Yale in 1934, joined the United States Navy during World War II, and served her country in uniform for over 42 years. Along the way, she became the first person in history to get a computer to understand English. So the story of machines that can speak, listen, and write—the story of Watson, Alexa, chatbots, Google Translate, and all the other linguistic marvels of the digital world—really all begins with Amazing Grace.

Grace Hopper, Queen of Software

Grace Hopper was born in New York City in 1906. As a young girl she learned quickly that her family held two values in especially high esteem: self-sufficiency and service to country. She and her parents were on a summer trip to New Hampshire, and Grace was out paddling a canoe by herself. Suddenly a gust of wind capsized the canoe, dumping Grace into the lake. Yet her mother, watching from the shore as Grace splashed about in the water, seemed strangely unconcerned. She merely grabbed a megaphone and shouted, "Remember your great-grandfather, the admiral!" Grace promptly swam back to shore, canoe in tow.[4]

The great-grandfather in question was Rear Admiral Alexander Wilson Russell, who fought the Barbary Pirates as a young man, and

* You can easily find her Letterman interview on YouTube.

who later served in the Union Navy. But Grace's military lineage went back even further. All her life she would tell the tale of Samuel Lemuel Fowler, an ancestor who'd marched with his musket to Concord, Massachusetts, in 1775 to stand up for his country. Grace would do the same 168 years later, though she stood up to fight alongside the British, rather than against them.[5]

In the fall of 1924, Grace Hopper shipped off to Vassar College, determined to prepare for the world of work. That very year, Vassar introduced three new courses to its catalog: "Motherhood," "Husband and Wife," and "The Family as an Economic Unit." Hopper didn't take those. She opted instead for "Electromagnetism," "Probability and Statistics," and "The Theory of Complex Variables"; with her mother's strong encouragement, she'd always been interested in math, and never in the traditional paths open to women of the day. At Vassar she flourished, graduating with honors in 1928 with degrees in math and physics, and soon thereafter headed to Yale to enter its PhD program in math.

In 1931, her dissertation not yet complete, Hopper returned to Vassar to join the math faculty, where her love and curiosity for the subject proved infectious. She became a wildly popular instructor; she took one course from an enrollment of 10 up to 75, with a waiting list on top of that. She had an unusual way of teaching math, short on abstraction and long on practical demos. She taught the mathematics of displacement, for example, by marching her entire class into a bathroom and having someone climb into a tub of water.[6] She finished her Yale dissertation from afar, graduating in 1934, and she continued teaching at Vassar for the next decade.

Hopper in the War: The Harvard Mark I

The outbreak of World War II would change Grace Hopper's life forever. In 1942, with the memory of Pearl Harbor—and her great-grandfather—fresh in her mind, she tried to enlist in the Naval Women's Reserve, one of the few military roles available to women. At 35, however, she was considered too old, and at 5′6″ and 105 pounds, 16 pounds below the weight requirement. She was rejected. But a fierce determination to serve was simply a given in Grace's family. She tried again, submitting special paperwork for an exemption to the weight requirement. This time she was accepted, and in December of 1943, she joined the U.S. Navy Reserves.[7] Midshipmen's school flew by quickly—as Grace put it, "Thirty days to learn how to take orders, and thirty days to learn how to give orders, and you were a Naval officer."[8] She graduated at the head of her class of 800 and was commissioned as a lieutenant (junior grade) in June of 1944.

Because of her math background, Hopper assumed that she would be assigned to a cryptography unit to help break the Axis powers' radio codes. As it turned out, however, there was something even more suited for a person of her background. She was told to report for duty in Cambridge, Massachusetts, where she would become only the third person to learn how to operate the Harvard Mark I, the country's first programmable digital computer. Years later, when Hopper was famous, and an interviewer asked how she got into computing, she said simply, "I was ordered to the first computer in the United States by the Navy, and I reported."[9]

The Mark I had been conceived by a man named Howard Aiken, built by IBM, and donated to Harvard—where Aiken was both a professor and a navy commander—in the name of war service. It was a beast of a thing, longer than a semitrailer and heavier than two rhinos: 5 tons, 51 feet long, 8 feet tall, and 3 feet deep. It had 530 miles of wire, 765,000 electromechanical switches, and a sleek modernist case designed by Norman Bel Geddes. The Mark I differed from other early computers in that it was truly all-purpose. It could handle differential equations, linear algebra, harmonic analysis, and statistics; it could be programmed to simulate a rocket, a submarine, a radar wave, you name it. Its cocreator Aiken called it "a general arithmetic machine," but the newspapers preferred terms like "robot brain" or "algebraic superbrain."[10] When military bigwigs came to visit, Aiken bragged that the Mark I was so fast that it could add three numbers every second, or do long division once every 14.7 seconds.[11] As an aside, it's interesting to compare these numbers with an iPhone X from 2017:

	Size (centimeters)	Weight (grams)	Additions per second
Harvard Mark I (1944)	1550 x 250 x 90	4,284,180	3
Apple iPhone X (2017)	14 x 7 x 1	138	350,000,000,000

Measured by computations per second per unit volume, an iPhone X is 4 million billion (4×10^{15}) times as capable. But the Mark I was still a lot faster at crunching numbers than a person. Besides, the line for the new iPhone was 73 years long.

When Hopper arrived at Harvard, her commanding officer gave her one week to learn how to program the Mark I. As you'll learn in a few pages, it was slow, frustrating work—yet there was no instruction manual, no tech-support chatbot, and no time for delay. It was the summer of 1944. Allied soldiers were storming the beaches of Normandy, and the Mark I team had been assigned to calculate the ballistics tables that showed the soldiers how to aim their new long-range artillery. Moreover, that was far from the team's only important project. The biggest one was "Problem K," in August of 1944: a highly classified and immensely complex set of calculations requested by a lab in Los Alamos, New Mexico. The Mark I was booked for weeks when this request arrived. But the navy said to give the Los Alamos mathematician—John von Neumann, working on something called the Manhattan Project—whatever time he asked for.

How Did You Talk to a Computer in 1944?

After the war, Hopper easily could have returned to Vassar and lived out her life as a full professor. She liked computers better than tenure, though, and so she remained in the Navy Reserves to begin her new life as one of the world's only computer experts. In that role, Hopper would soon take a historic step, borne of the frustration she experienced programming the Mark I: she would become the first person ever to talk to a computer in English.

Hopper worked at Harvard's Computation Lab for four years after the war. Then in 1949 she accepted a job at the Eckert–Mauchly Computer Corporation, which made a computer called the UNIVAC. It was a fateful decision, both for Grace and for the future of computing. Before the UNIVAC, computers were seen as great tools for doing calculations in math and science, but not much else. Experts thought that there might be a demand for 20 computers in the whole country, mostly at government research labs. But Hopper's work on the UNIVAC changed that. She demonstrated that computers were also useful for solving business problems involving *databases*—a complication that simply never arose in the pure math problems that the Mark I had solved during the war. With Hopper's help, companies everywhere began to see the potential in these new machines. U.S. Steel ended up buying a UNIVAC to run its payroll. MetLife bought one to calculate insurance premiums. DuPont, General Electric, the Census Bureau, Westinghouse—they all bought a UNIVAC to crunch their data, making it the world's first commercially successful computer.[12]

To explain Hopper's big breakthrough here, we need to return to a question from 1944: How did Hopper give instructions to the Harvard Mark I? How did she tell 765,000 electromechanical switches to dance a tune that calculated a ballistics table?

It certainly wasn't in English. Hopper described it like this: "You simply broke down all your processes of mathematics into a series of very small steps of add, multiply, divide . . . and you put them in sequence."[13] She made it sound so simple. It wasn't. The hard part was phrasing these instructions in the Mark I's own "machine language," the only language it could comprehend.

To understand what a machine language is, imagine a computer program for making tea. In a modern "high-level" programming language like Python, the program might read like this: (1) Put 2 teaspoons of tea in a teapot. (2) Boil 16 ounces of water. (3) Pour the boiling water over the tea and let it steep for 4 minutes. But in a machine language, you'd have to break those instructions down into much smaller, more specific tasks. Instead of saying "boil water," you'd first describe how to walk to the sink: move your left foot, move your right foot, move your left foot, again and again. Next you'd describe how to fill the kettle: raise your left hand, grasp the faucet, twist the handle counterclockwise, and so on. Then you'd describe how to heat the water, brew the tea, and pour the tea, all in similarly tedious biomechanical detail. Moreover, you'd have to issue each instruction not in English but using numeric codes fed into the computer by punching holes in a long tape of perforated paper. These codes told the Mark I exactly how to manipulate the bits ("binary digits," or 0s and 1s) in its internal circuits. As the programmer, you had to know which codes did which things. In our tea-making example, code 72 04 might mean "move left foot," 61 07 might mean "grasp faucet in left hand," and so on.

So: 72 04, 61 07 . . . that's typical machine language. It's a long way from "To be, or not to be"—a long way, even, from "Alexa, play me some eighties music." As author Douglas Hofstadter put it, "Looking at a program written in machine language is vaguely comparable to looking at a DNA molecule atom by atom."[14] But this is how computers "think," even today. And at the dawn of the digital age, there was simply no other way to tell them what to do. As a programmer of that era, you were basically a binary plumber, piping bits through a computer's circuits with the help of a codebook that showed how to translate math problems into machine language. You would look up items in the codebook, punch the right holes in the tape, feed the tape into the computer, and cross your fingers.

Talking to a computer like this was tedious and error-prone. Worse still, some early computers didn't even use ordinary decimal (base 10) numbers. Instead they used octal (base 8), which led to mind-bending arithmetic like $7 + 1 = 10$ or $5 \times 5 = 31$.[†] This drove programmers nuts, and Grace Hopper was no exception. Once, after spending weeks programming in octal on a computer called the BINAC, Hopper realized that she couldn't get her checkbook to balance. She redid the numbers again and again, but she couldn't find her mistake. Then it hit her. Her numbers didn't match the bank's because she'd been keeping her checkbook in octal without even realizing it.[15]

† The ordinary decimal numbers 8 and 25 are expressed in octal as 10 and 31, respectively.

Hopper Invents the Compiler

To Hopper, the checkbook episode really drove home the problem with computers: they just didn't speak our language. But she also saw the possibility of a better way. It all started with the idea of noting and recording common patterns, or what later came to be known as a "subroutine."

Maybe you know the story about the prisoners who've heard every joke so many times that they've given each one a number, just to make things easier. So one guy yells out "31!" and the other prisoners burst into laughter. Another guy yells "17!" and there's even more laughter. Then a third guy yells out "104!" but this time everybody goes silent, because the humor is all in how you tell it.

In computing, a subroutine is like a numbered joke: a generically useful piece of code for solving a problem that comes up again and again, like factoring a quadratic equation or sorting a list of numbers. Whenever Hopper wrote a subroutine for the Mark I, she would copy it down by hand in a notebook, so she wouldn't have to reinvent the wheel next time. Before long, she had a big collection of subroutines written in machine language. When she wanted to use one again, she would copy it back out of her notebook onto the tape. This took a long time, and a single copying mistake would ruin the whole program. Hopper realized, however, that the Mark I was way better at copying than any human. This gave her an idea. Why not store a library of these subroutines—indexed by mathematical code words, just like the numbers on the prisoners' jokes—and program the computer to copy and compile whatever subroutines were needed for a given task? In other words, why not program the computer to program itself?

This idea of a "compiler" is perhaps the most important software innovation in computing history. To return to the tea-making metaphor, programmers no longer had to code up instructions like "raise left hand, grasp faucet, turn faucet" whenever they wanted to fill the kettle. Instead, they could just give the machine the right high-level code words to fill the kettle, boil the water, etc. The machine itself would then compile all the right subroutines into a program for making tea. Programs that used to take a week to write would now take five minutes. Moreover, each subroutine was debugged in advance and known to work. You couldn't possibly tell the joke wrong.

When Hopper first explained this idea to her bosses, they thought she was crazy. Computers could only do math, she was told. They couldn't possibly write their own programs; only a human could do that. It was around this time that Grace picked up her favorite remark, which she would repeat many times over the years: "The most dangerous phrase in the language is, 'We've always done it that way.'" Of course Hopper's bosses were wrong, as she would eventually prove.[16]

Hopper didn't stop there. Her insight about compilers convinced her of one essential truth: that the future of computers depended on making them easier to talk to. That required much more than just substituting mathematical code words for the old binary-plumbing manual. Math notation was fine for a scientist or a navy researcher, but most potential computer buyers, Hopper observed, "wouldn't know a cosine if they met one walking down the street."[17] Business people didn't need a language for telling a computer how to calculate the trajectory of a rocket. They needed a language for working with databases: accounts, prices, sales, payroll, hours worked, and so on. There was only one common language for talking about data in a way that cut across different areas of commerce: English. To Hopper, the conclusion was obvious. Computers needed to be programmed to work with English inputs.

Once again, her bosses thought it was foolish even to try, and in 1953 they denied funding for her proposal. Of course we can't get a computer to understand English, they told her. The whole idea is absurd. We have to program computers using symbols and math. We've always done it that way.[18]

It wasn't the first time she'd battled computing's macho culture, and it wouldn't be the last. But when the wind dumped her out of the canoe, Grace just swam back to shore. She pursued her idea on the side, and by January of 1955 she had a working prototype. Speaking to a room full of top company executives, she demonstrated that her "data-processing compiler" could make the UNIVAC understand an English-language program whose first several lines looked like this.[19]

```
Input Inventory File A; Price File B;
Compare Product #A With Product #B.
If Greater, Go To Operation 10;
If Equal Go To Operation 5;
```

And so on. Hopper had programmed the computer to translate these phrases behind the scenes, allowing users to focus on what they knew (the flow of data) rather than what they didn't (the details of the math).

Then Hopper made a misstep. To emphasize that the computer was just applying rules to match phrases with bit patterns, she demonstrated that the equivalent phrases in French could yield the same program: "Lisez-paquet A; Si Fin de Donnes Allez en Operation 14." This sent her bosses into a tizzy. As Grace described: "That hit the fan! It was absolutely obvious that a respectable American computer, built in Philadelphia, Pennsylvania, could not possibly understand French."[20] With a few suspiciously un-American phrases, she'd set the project back four months.

Eventually Hopper prevailed, and she won company funding to develop her data-processing compiler, called FLOW-MATIC. A pilot study showed that FLOW-MATIC was letting customers accomplish the same tasks as the old "math and symbols" method, but in a quarter of the time. With their customers so gung-ho about Hopper's new approach, the bosses had no choice but to relent. They were lucky Grace had been so determined.

With that, the Programming Language Revolution had begun. From the mid-1950s onward, virtually everyone who talked to a computer did so using the model that Grace Hopper pioneered. Machine languages remained important, but they became the province of a few highly trained specialists. Everyone else used high-level programming languages whose instruction sets were much more similar to "fill the kettle" than to "grasp the faucet, twist the faucet." In establishing this model, Hopper had an enormous impact on what we believe is the most important trend in human history since 1945: the spread of digital technology into every part of life.

From Grace to Alexa: The Natural Language Revolution

That brings us to about 1960. So how did we subsequently get to the point where you can have any consumer good in the world delivered to your door, simply by speaking your request out loud to a computer?

To set things up, let's summarize the top-down, rules-based model for human-machine linguistic interaction that Grace Hopper pioneered in the 1950s.

- People tell machines what to do using a programming language, with a heavily restricted grammar and a small vocabulary of English words.
- Machines interpret these commands in their own language using huge books of preprogrammed translation rules, operating behind the scenes.
- Both the programming rules for humans and the translation rules for machines have to be defined from scratch, one by one, by human programmers.

From the 1950s through the 1970s, experts tried to get machines to understand natural language using this same top-down approach: (1) place constraints on human users, by restricting the grammar and vocabulary they can use; and (2) program the machines chock-full of translation rules: syntax, pronunciation, word choice . . . basically, all the rules you learned without trying as a child, together with all the grammar rules you learned from Mrs. Thistlebum in elementary school.

This rules-based philosophy had worked great for programming languages. But it never worked very well for natural languages.

A great example of how it went wrong is computer speech recognition. The very first speech-recognition systems were essentially toys. At the 1962 World's Fair, for example, IBM showed off a machine that could recognize spoken English words—precisely 16 of them, and only if enunciated with painful clarity. In the 1970s there was a false dawn, in the form of a program called Harpy, created by researchers at Carnegie Mellon. Harpy recognized exactly 1,011 words, about as many as a small toddler. It was built on Grace Hopper principles: a restricted grammar and vocabulary for humans, and a fiendishly complicated set of rules for the machine, to transcribe speech into text. Harpy's team of five programmers spent two full years coding up these rules—rules for acoustics, phonetics, sentence structure, word boundaries, and so on. In highly idealized laboratory conditions, the system even reached 70% word-level transcription accuracy. This caused a lot of excitement among AI researchers. Harpy seemed to suggest that, with better rules and faster computers, human-level performance might be just around the corner.[21]

Yet these hoped-for improvements in speech recognition never materialized. In later tests involving real-world conditions, Harpy's word-level accuracy fell to 37%. After five years, the U.S. government cut funding for the project. And today, pure rules-based systems for natural language processing have become vanishingly rare. In the end, they were never able to overcome three basic problems: rule bloat, robustness, and ambiguity.

Problem 1: Rule Bloat

First, it's really hard to write down all the rules for natural languages. There are way too many of them, vastly more than any programming language. Although you may not know it, you can actually learn a lot of Python in a day. But you can't learn a lot of Korean in a day.

Part of the problem is that all rules have exceptions—or as famous linguist Edward Sapir put it, "All grammars leak." In English, for example:

- Adjectives come before nouns, but don't tell that to the attorney general's heir apparent.
- "*I* before *e,* except after *c,*" except for weirdly prescient scientists who drink protein shakes with caffeine.
- Two positives don't make a negative? Yeah, right.

And so on. Exceptions create trouble, since machines are nothing if not tyrannically consistent about the rules. The only way to handle this is to write a rule for every exception.

This sounds painful enough, but the "too many rules" problem runs much deeper. For many aspects of language, we simply don't know what the rules are. Consider what linguists refer to as the "speech segmentation" problem. Try reading the following phrase aloud: "The weather report calls for rain tomorrow." You will perceive this as a discrete series of words: "weather," "rain," "tomorrow," etc. But this discreteness is a cognitive illusion; only sci-fi robots . . . speak . . . with . . . pauses. What you're really hearing is a continuous stream of sound, with no acoustically obvious gaps between the words. Figuring out where one word ends and the next begins is a really hard problem. Linguists have discovered all sorts of hidden auditory rules that we rely upon to do this, with names like "phonotactics" and "allophonic variation." But linguists also know that they haven't found *all* the rules, because the ones they *have* found can't possibly explain how good we are at speech segmentation.

The problem is obvious: if you can't even identify all the rules, then you certainly can't teach them to a computer.

Problem 2: Robustness

The second problem with top-down rules is that they usually break upon contact with the real world. Put simply, they aren't robust.

Consider, for example, the problem of distinguishing speech from background noise. Your brain is *incredible* at this; you can usually understand your friends in a noisy bar, despite the clatter. Neuroscientists don't 100% understand how you do this, which is why background noise is a perennial complaint for people who wear hearing aids.

Another source of irrelevant variation is what you might uncharitably call "mistakes." Imagine what would happen if you disregarded one rule of English in every sentence you wrote or spoke today. You might get a few odd looks, but people would understand you just fine. Even if you use sentence fragments. Even if like Yoda you speak. Even if you let your modifiers dangle, unmoved by Mrs. Thistlebum's contempt for your loutish abuse of participles. People's understanding of language is highly robust to this kind of variation, but that robustness is hard to replicate using top-down rules.

Another big issue is pronunciation. Ask someone from Derry, New Hampshire, to say "caramel." Then ask someone from Derry, Northern Ireland. The answers will sound nothing alike; you won't get the same vowels, not even the same number of syllables. You might respond: OK, let's just make two rules for "caramel." But that's just the Northern Irish and the Yankees. Don't forget the Texans and the Londoners and the Californians and . . . well, you can see the problem. It's rule bloat all over again. Imagine trying to program a computer with a suite of rules that will robustly map all these different pronunciations—"care-a-mell," "crrr-mul," and every shade of caramel in between—to the same underlying word. Congratulations, you've just solved speech recognition for "caramel." Only 171,475 words in the *Oxford English Dictionary* to go. Then you can start on slang, and maybe then Mandarin.

Problem 3: Ambiguity

Finally, it's hard to come up with rules that are good at handling ambiguity—and language is full of ambiguity. The most obvious examples involve homophones: weather/whether, rain/reign, I scream for ice cream, and so on. Then there's what linguists call "syntactic ambiguity," or sentences that can be parsed in multiple ways. Newspaper headlines are often guilty of this, yet we can usually understand them:

- DEFENDANT GETS NINE MONTHS IN VIOLIN CASE is not about an unusual form of confinement.
- INCLUDE YOUR CHILDREN WHEN BAKING COOKIES is not a recipe suggestion.
- BRITISH LEFT WAFFLES ON FALKLANDS is about an indecisive Labour Party, not an abandoned breakfast.

The basic issue here is that when it comes to language, you are an incredible probabilistic inference engine, programmed by eons of evolution to scoff at ambiguity. You barely notice the mssng vwls in text messages. You cut through analogies like a hot knife through butter. You know that "we need to take a break" means something different in a basketball game than in an argument with your partner. You use context information to interpret what someone meant to say, even if there's another plausible sentence that sounds exactly the same:

- "The president's new direction has split his party."
- "The president's nude erection has split his party."

These kinds of ambiguities are defined out of existence by any language designed for computers, precisely because of the trouble they create for a rule-crafter. Yet you seem to handle them with little trouble. How?

1980–2010: The Growth of Statistical Natural Language Processing

Philosophers like to distinguish between two kinds of knowledge: knowing how versus knowing that. "Knowing how" means intuitive, practical knowledge. For example, you know *how* to walk and how to ride a bike without conscious effort. "Knowing that," on the other hand, means factual, textbook knowledge. For example, from reading a random page on Wikipedia beginning with the letter *N*, you could know *that* Nike is a brand of shoe, or that Napoleon invaded Russia in 1812 but found it rather cold.

For people, spoken language seems like the ultimate example of "knowing how." The cognitive miracle is how effortless it all seems—how we can speak intelligibly, and decode the ambiguous sound waves emanating from other people's mouths, without even thinking about it. To do this, you automatically draw inferences using all kinds of side information: your experience of the world, your implicit understanding of what other people are likely to be thinking, and lots of subtle auditory clues.

As we've seen, NLP experts spent a long time trying to get computers to understand natural language by giving them lots and lots of explicit rules. These rules were supposed to mimic the know-how that children pick up naturally when they learn language. Yet even at their best, these rules-based approaches made ghastly errors. They couldn't match the linguistic skill of a typical five-year-old, much less an adult.

After three decades of experience, it had become obvious to experts in natural language processing that a new approach was necessary. This approach would have to be flexible rather than rigid. Probabilistic rather than deterministic. Bottom-up, based on real-world data, rather than top-down, based on a profusion of rules. Above all, it would have to handle the way people really talk, rather than how a grammarian thinks they ought to.

So in the 1980s, researchers tried something different. They threw up their hands, threw out the rules, and said: let's just use data. They invented new algorithms that worked on a very different premise: human linguistic knowledge may be too hard to reverse engineer, but this knowledge *does* have a clear statistical shadow, discernible in how we speak and write. For example, if "weather report" makes more sense than "whether report," then a vast collection of real sentences should contain many more examples of the former than the latter. Sure enough, that's just what we see. We used an online tool called the Google Ngram Viewer,* which lets you track the popularity of any word or short phrase across all published books in English. We learned that between 1950 and 2000, about 150 out of every billion two-word phrases in published books were "weather report" (0.0000155797%). This is about 250 times more common than "whether report" (0.0000000652%), which is used mainly as a bad pun or an example of phonetic ambiguity.

From the 1980s onward, NLP researchers began to recognize the value of this purely statistical information. Before, they'd been hand-building rules capable of *describing how* a given linguistic task should be performed. Now, these experts started training statistical models capable of *predicting that* a person would perform a task in a certain way. As a field, NLP shifted its focus from understanding to mimicry—from knowing how, to knowing that.

These new models required lots of data. You fed the machine as many examples as you could find of how humans use language, and you programmed the machine to use the rules of probability to find patterns in those examples. Language became a prediction-rule problem based

---

* https://books.google.com/ngrams. An "n-gram" is a geeky linguistics term for a phrase containing *n* items, like words or symbols. For example, "weather report" is a 2-gram, since it contains two words.

on input/output pairs, similar to the problems solved by Henrietta Leavitt, or that farmer in Japan who uses deep learning to classify cucumbers:

- For speech recognition, you pair a voice recording (input = "ahstinbrekfustahkoz") with the correct transcription (output = "Austin breakfast tacos").
- For translating English to Russian, you pair an English word or sentence ("reset") with the correct Russian translation ("perezagruzka").
- For predicting sentiment, you pair a sentence ("What a delightful morning spent in line at the DMV") with a human annotation (☹).

And so on. In each case, the machine must use the data to learn a prediction rule that correctly maps inputs to outputs.

In the 1980s, speech-recognition software based on this principle began to hit the market. These systems could recognize a few thousand words, but only if you spoke . . . like . . . a . . . robot. The 1990s and early 2000s saw richer models that performed incrementally better and let you speak at a natural pace. But the huge bottleneck here was the availability of data. You may recall the "overfitting" problem we described in chapter 2, in which a complicated model simply memorizes the random noise in a small data set, without learning the underlying pattern. It was the same problem here. Researchers in NLP simply didn't have enough data to construct models that were sufficiently complicated to describe human language without overfitting what little data they had. As a result, by the 2000s, speech recognition again hit a plateau, at about 75–80% word-level accuracy. For nearly a decade, progress was discouragingly slow—and not just for speech recognition but also for other tasks in natural language processing that were hampered by a lack of data, from machine translation to sentiment analysis.

Post 2010: The Natural Language Revolution

Around 2010, everything started to change—slowly at first, then at a startling pace. What drove this change was a massive infusion of data.

Jorge Luis Borges once wrote a story called "The Library of Babel," about a library whose books contained all possible works of prose: that is, all possible orderings of the letters of the alphabet and the basic punctuation marks. Most books in this library were pure monkey-at-a-typewriter nonsense, but somewhere, in some book, you could find every possible sentence—every love story, every adventure, every work of genius ever written, or that might be written.

Our real-life Library of Babel is called the internet, and while we're not quite in Borges territory yet, we're getting closer. Think of the immense collections of spoken and written English sentences found on the servers of the world's major tech firms. Think of a library with every book, magazine, newspaper, journal, song, film, and play ever written. Now think *way* bigger. Every web page. Every email in history. Every Google search or product review, every text message ever sent, every chat on Slack or Skype, every post on Facebook or Twitter, every comment on YouTube or Instagram. By volume, this collection of sentences makes the Library of Congress look like a third-rate bookmobile. And around 2010, the best minds in AI had finally developed the right tools for using all that data to its full effect.

Some of this data simply landed in the laps of the big tech firms, but these firms also went far out of their way to collect more. One example was Google 411, which debuted in 2007. You may remember a time when people dialed 411 to look up a phone number for a local business, at a dollar or so per call. Google 411 lets you do the same thing for free, by dialing 1-800-GOOG-411. It was a useful service in an age before ubiquitous smartphones—and also a great way for Google to build up an enormous database of voice queries that would help train its statistical models for speech recognition. The system quietly shut down in 2010, presumably because Google had all the data it needed.

Of course, there's been an awful lot of Grace Hopper–style coding since 2007 to turn all that data into good prediction rules. So more than a decade later, what's the result? Let's try a simple experiment. Open up a blank email on your phone and try dictating a test phrase: "The weather report calls for rain, whether or not the reigning queen has an umbrella." If you're a native English speaker and your phone runs iOS or Android, it will almost surely get the sentence right, without confusing weather/whether or rain/reign.

That's one tiny little example of what data buys you. The software knows that "whether" and "reign" are statistically more likely in some contexts, while "weather" and "rain" are more likely in others. This isn't because your phone somehow understands the *meaning* of words. There's no meaning involved, just a rich set of context-specific probabilities[‡] for basically every English word and phrase ever uttered on the internet. When the acoustical data is ambiguous, your phone breaks the tie using these probabilities. While certain things still give it trouble, at least in 2018, the software is getting better all the time.

Other NLP systems have improved rapidly, too, and for the same reason. Take machine translation. For many years, there was a cottage industry of internet memes devoted to errors made by Google Translate, of which you can find hundreds scattered across the web. For example, some wise guy realized back in 2011 that when translating from English to Vietnamese, "Will Justin Bieber ever reach puberty?" became "Justin Bieber will never reach puberty."[22] This kind of syntactical error was a classic failure mode of older machine-translation algorithms. They'd get the words mostly right, but they'd often bungle the word order in the target language, producing something that was either wrong or nonsensical.

As more training data has become available, and as the statistical models of language have gotten better, these gross syntactical errors have gotten much less frequent.[23] Moreover, all that data has made explicit translation rules much less important. For example, nobody explicitly told Google Translate that English uses subject-verb-object order, as in "programmers love coffee," but that Japanese uses subject-object-verb order, as in "programmers coffee love." The algorithm simply learned syntax from statistics—from millions of sentences of training data, rendered side by side in English and Japanese.

Today, the vast majority of successful NLP systems represent the ultimate example of the philosophers' second kind of knowledge: knowing that, as opposed to knowing how.[24] To the software, it's all just

‡ Here "context" just means "the other words in the sentence."

facts. But these facts are usually enough, because the data sets themselves are so big, and the algorithms so sophisticated.

How Words Become Numbers

So let's now turn to the question of algorithms. Suppose you have a giant database of sentences—a Library of Babel for 100+ natural languages, from English to Chinese to Farsi. How do you build an AI system for a language-related task and actually make it work?

As you might imagine, there are a lot of details that we can't go into here, because they're just too complex. Those details are the reason a company like Google has 70,000 employees, a small army of PhDs, and more computers than you've ever seen in your life. We can, however, give you a high-level explanation of something really important here, called a "word vector." Specifically, we're going to explain Google's famous word2vec model, which provides a numerical ("vector") description for every word in English. If you understand word2vec, then you understand one of the most fundamental AI ideas of the last decade. Even systems that don't use this algorithm directly still use the same underlying approach.

Word2vec answers a simple question: How do we turn words into numbers, so that words with a similar meaning have similar numbers? This may sound bizarre or even impossible. In what sense can the meaning of words like "toaster" or "courage," or a phrase like "Toronto Maple Leafs," be described as numbers? But we claim that it's not nearly as hard as it sounds. In fact, children do it all the time.

The Math of 20 Questions

There's a scene in *A Christmas Carol* that takes place in the living room of Ebenezer Scrooge's nephew Fred. Fred had invited his wealthy skinflint of an uncle to Christmas dinner, only to be told that "every idiot who goes about with 'Merry Christmas' on his lips should be boiled with his own pudding, and buried with a stake of holly through his heart." But in the meantime, Scrooge has been visited by spirits who show him the error of his miserly ways, and the third spirit, the Ghost of Christmas Present, has brought Scrooge to Fred's house on Christmas Day. They look on, unseen, as Fred and his family play a game called Yes and No. Scrooge's nephew Fred has to think of something, and everyone else in the room has to find out what it is by asking only yes-or-no questions:

> The brisk fire of questioning to which he was exposed, elicited from [Fred] that he was thinking of an animal, a live animal, rather a disagreeable animal, a savage animal, an animal that growled and grunted sometimes, and talked sometimes, and lived in London, and walked about the streets, and wasn't made a show of, and wasn't led by anybody, and didn't live in a menagerie. . . .

What could it be? The children are now in conniptions of laughter. They make several guesses, learning that Fred wasn't thinking of a bear or a horse, nor a tiger or a donkey. Then at last Fred's sister-in-law comes up with the answer: "I know what it is, Fred! I know what it is! . . . It's your Uncle Scro-o-o-o-oge!" And it was.

American kids grow up calling this game 20 Questions, and although it may not look like it, it's a very mathematical game. In fact, 20 Questions shows us how to turn words into numbers, just like AI systems do. Let's take the word "Scrooge" from the game in Fred's living room. Its numerical representation looks like this:

	Animal	Agreeable	Growls or grunts	Talks	Lives in London	Is a bear
Scrooge	1	0	1	1	1	0

This is called a "word vector."[§] Specifically, it's a "binary," or 0/1 vector: 1 means yes, 0 means no. Different words or phrases, like "Tiny Tim" or "Paddington Bear," would produce different answers to the same questions, so they would have different word vectors. If we stack all these vectors in a matrix, where each row is a word and each column is a question, we get something like this:

§ In math, a vector is just a set of numbers all associated with the same thing.

	Animal	Agreeable	Growls or grunts	Talks	Lives in London	Is a bear
Scrooge	1	0	1	1	1	0
Rafael Nadal	1	1	1	1	0	0
Tiny Tim	1	1	0	1	1	0
Paddington Bear	1	1	0	1	1	1
Trafalgar Square Christmas tree	0	1	0	0	1	0

How AI Plays 20 Questions

So it's actually pretty easy to turn words into numbers, using a game of 20 Questions. Let's now change the rules in three ways—both to make it much more like the game an AI system must play, and to make the word vectors we get out as packed full of meaning as possible.

First rule change: you don't merely win or lose. Instead, you get scored by "semantic closeness," or how close you are to the word's underlying meaning. Let's not get into the details of what "close" means. In AI there's a detailed mathematical answer, but you should just imagine the most fair-minded person you know acting as judge. For example, suppose the answer is "bear":

- If your final guess is a bear, you get 100 points.
- If you guess a dog or a wolverine, you might get 90 points. You're pretty close, phylogenetically speaking.
- If you guess a mosquito, you might get 50 points. At least you guessed another animal.
- If you guess cough syrup, you get 2 points. You're way off, but coughing might remind you of a growling bear in some tiny way.

This way of keeping score matches the real-world design requirements of most NLP systems. For example, if you translate JFK saying "Ich bin ein Berliner" as "I am a German," you're wrong, but a lot closer than if you translate it as "I am a cronut."

The second rule change is that instead of just a yes or no, each answer is a number between 0 (completely no) and 1 (completely yes). For example, take "Is it a bear?"

- For an actual live bear, you'd answer with a 1.
- For a talking bear like Paddington, you might answer with a 0.9. He's still a bear, but not quite a Platonic-form-of-bear.
- For Scrooge, you might answer 0.65. He's not really a bear, but he does have strong bearlike tendencies. (In *A Christmas Carol,* Fred's family even complained "that the reply to 'Is it a bear?' ought to have been 'Yes'; inasmuch as an answer in the negative was sufficient to have diverted their thoughts from Mr. Scrooge.")
- For Rafael Nadal, you might answer 0.2. He seems to be a lovely nonbearlike person on TV, but he is alive, and he does grunt a lot.

With this rule change, we have word vectors made of continuous numbers—0 *through* 1 for each question, rather than 0 *or* 1; gray, rather than black or white.

Now for the biggest rule change: you must ask the *same questions in every game.* This would surely kill the fun at your next Victorian-parlor-games party, since it reduces every game of 20 Questions to Let's Play Census, where you fill out the same dreary survey form. Nonetheless, put your qualms aside. Imagine trying to come up with questions that are both broad and rich enough to distinguish every possible word, from "Scrooge" to "screwdriver," from "barbecue" to "basketball," from "erythrocyte" to "epistemology."

Not easy, right? But this is how models of natural language in AI actually work. One caveat: we're going to need many more than 20 questions, since we can no longer adapt our later questions to the earlier answers. So in AI, we play 300 Questions instead.

As far as what questions to ask, we won't address that directly. Instead, let's talk about process. Your first thought might be to do this by committee: put some smart people in a room and tell them they can't leave until they come up with 300 questions that collectively provide a unique encoding for every word or phrase in English. This might be an interesting Biodome-style sociological experiment. But if you think it's likely to work, then you have a lot more faith in committees than we do. At the very least, it would take a very, very long time. And we want to tell our mobile phones to order us a pizza right now, not when some committee makes up its mind.

That leaves us with only one good way to write questions: let an algorithm choose.

What kind of questions could an algorithm even ask? Questions about meaning won't work, because machines don't understand meaning. But they do understand *word co-location statistics*—that is, which words tend to appear with which other words in real sentences written by humans. These statistics are surprisingly good surrogates for meaning. Here's an example question along these lines: "Take all sentences with 'fries,' 'ketchup,' or 'bun' in them. Does this word also appear frequently in those sentences?" To be sure, this is the kind of question that a neural network might have Ross ask in a new episode of *Friends*. Crucially, though, it's also a question that a machine can both pose and answer, because it doesn't require *understanding*, merely counting.

Of course, this particular question is too narrow if you're allowed only 300 of them. But the basic premise—asking questions about word co-location statistics—is sound. While we're omitting lots of details here, that's basically what word2vec does. Through trial and error, it learns 300 good sets of probe words (analogous to "bun" and "ketchup" in the example above).[25] It then plays 300 Questions over and over again, learning a word vector for each word or phrase in English based on its co-location statistics with the probe words.

The resulting word vectors can be stacked in a matrix, one word per row, just like we did for "Scrooge" and "Tiny Tim" a few pages ago. This is an enormous matrix, with 300 columns and millions of rows. On the next page, we've given you a subset of 4 columns and 40 rows, to give you a sense of the questions that the algorithm learns to ask.

HOW AN ALGORITHM PLAYS 20 QUESTIONS

	Question 1: "computers"	Question 2: "universities"	Question 3: "cooking"	Question 4: "law"
nvidia	1	0.045	0.156	0.083
servers	0.999	0.944	0.214	0.184
username	0.999	0.468	0.842	0.963
ethernet	0.999	0.587	0.617	0.072
interface	0.999	0.355	0.831	0.032
router	0.998	0.697	0.986	0.911
displays	0.998	0.693	0.111	0.174
port	0.997	0.646	0.583	0.184
pixels	0.997	0.253	0.017	0.21
firewall	0.995	0.729	0.957	0.636
undergraduate	0.089	0.999	0.107	0.627
faculty	0.365	0.999	0.114	0.944
scholarships	0.063	0.999	0.291	0.398
applicants	0.153	0.999	0.22	0.77
colleges	0.206	0.997	0.132	0.514
fellowship	0.216	0.997	0.035	0.688
committee	0.32	0.996	0.912	0.824
departments	0.42	0.994	0.502	0.77
residential	0.145	0.993	0.569	0.801
publications	0.173	0.993	0.524	0.938
roasted	0.778	0	1	0.767
smoked	0.596	0.012	1	0.799
beers	0.815	0.043	1	0.613
bbq	0.182	0.077	1	0.039
corn	0.827	0.044	1	0.122
beef	0.471	0.015	0.999	0.699
chili	0.403	0.002	0.999	0.425
peppers	0.398	0	0.999	0.572
grilled	0.531	0.001	0.999	0.46
flavor	0.281	0.026	0.997	0.248
bail	0.221	0.63	0.923	1
custody	0.509	0.536	0.943	1
arrest	0.149	0.444	0.839	1
charges	0.002	0.157	0.57	1
penalties	0.44	0.105	0.413	0.999
possession	0.123	0.304	0.73	0.999
illegal	0.045	0.406	0.478	0.999
conviction	0.015	0.121	0.928	0.999
lawsuit	0.175	0.147	0.735	0.999
sheriff	0.275	0.305	0.882	0.999

For example, in the first column, we see words like "router," "pixels," and "firewall," whose answers to question 1 are all very close to 1. It's clear that the algorithm has learned to ask a question along the lines of "Does the word tend to appear in sentences with computer words?"* (Remember that under our revised rules, 1 means "completely yes" and 0 means "completely no.") Similarly, in the third column, we see words like "roasted," "smoked," "beef," and "grilled," all with answers near 1. The algorithm must have learned to ask a question about cooking with fire. It's also learned to ask questions about universities and criminal law—and, in other columns not shown, questions about animals, politics, sports, health, and hundreds of other topics.

Putting Word Vectors to Work

This approach provides a surprisingly nuanced picture of language. AI researchers even have a favorite parlor trick to show off the richness of their word vectors: answering SAT-style analogy questions, merely by addition and subtraction. For example, take the analogy "man is to king, as woman is to _____?" How might we phrase this analogy as a math problem, suitable for describing in terms of arithmetic on word vectors?

Here's how. Take the vector for the word "king" and *subtract* the vector for the word "man." (We can add and subtract vectors just like numbers, because vectors *are* numbers.) Intuitively, by subtracting "man" from "king," we've stripped the word "king" of its male-gender component, resulting in a new vector that presumably represents a gender-neutral concept of royalty. To this new vector, now *add* the vector for the word "woman," thereby mathematically reintroducing a gender component—this time a female one. In other words, *take the word "king" and make it female,* which we express in terms of arithmetic as "king – man + woman." Word2vec's answer is spot on: if you actually do the arithmetic, you get the word vector for "queen."

---

* Of course, the algorithm doesn't *know* that the question is "about computers." It merely knows that the question is about co-location statistics involving other words that we, as humans, can subsequently interpret as being about computers.

Other kinds of analogies work the same way, using vector addition and subtraction.

- World capitals: London – England + Italy. Word2vec's answer: "Rome."
- Word tenses: captured – capture + go. Word2vec's answer: "went."
- Which hockey teams play in which cities: Canadiens – Montreal + Toronto. Word2vec's answer: "Maple Leafs."

In effect, word2vec has learned to take the SAT Verbal test using only skills from the SAT Math test. The underlying model knows absolutely nothing about monarchy, gender, geography, grammar, hockey, or anything real about the world. It knows only about the statistical properties of word usage gleaned from training data, together with the rules of probability.[26]

This might be a parlor trick, or just a delightfully geeky diversion for coders, but it also emphasizes an important point: once you've turned words into vectors, you can do math with them. This is essential to building AI systems for language. Computers don't understand words, but they do understand math.

Take speech-recognition software, like the kind that powers Alexa or Google Voice. Word vectors help tremendously here, because they encode the context of a sentence in a mathematical language that a computer can work with. Among other things, this provides crucial tie-breaking information for homophones, like "rows" versus "rose." Even though these words sound alike, they have different word vectors—different answers in AI's big game of 20 Questions—and one of these vectors will usually fit in better with the vectors of the surrounding words in any particular sentence. Admittedly, what it means to "fit in better" is very complicated, and not worth going into here. It boils down to a fancy calculation involving vector arithmetic, which produces probabilities that break the tie when the acoustical information is ambiguous, like this:

How	lovely,	it	smells	like	a	[skunk /	rows /	**rose** /	goat /	sewer ...]
0.45	0.61	0.83	0.39	0.45	0.43	0.66	0.22	**0.71**	0.32	0.20
0.37	0.18	0.51	0.39	0.71	0.98	0.22	0.31	**0.48**	0.87	0.26
...	...	...	...	...	...	...	...	**...**	...	...
0.99	0.33	0.39	0.24	0.29	0.19	0.68	0.71	**0.30**	0.26	0.06

Or this:

On	the	third	day,	he	[rows /	**rose** /	telephoned...]	from	the	dead.
0.46	0.40	0.47	0.59	0.35	0.22	**0.71**	0.26	0.24	0.40	0.16
0.62	0.68	0.93	0.77	0.41	0.31	**0.48**	0.62	0.34	0.68	0.66
...	...	...	...	...	...	**...**	...	...	...	...
0.27	0.63	0.85	0.43	0.94	0.71	**0.30**	0.30	0.28	0.63	0.10

Or this:

He	planted	100	[ears /	**rows** /	rose ...]	of	corn.
0.35	0.75	0.37	0.19	**0.22**	0.71	0.83	0.45
0.41	0.75	0.23	0.22	**0.31**	0.48	0.75	0.15
...	...	...	...	**...**	...	...	...
0.94	0.25	0.80	0.96	**0.71**	0.30	0.04	0.99

Word vectors provide a clear-cut mathematical description of something that to any human listener seems simple: sometimes one word fits better, and sometimes the other. And with some important task-specific modifications, the same basic math powers machine translation, chatbots, search-by-voice systems . . . even neural networks that can write about baseball.

Humans and Machines, Talking Together

We hope that you now understand some of the key ideas that have led us to the present moment, when machines have clearly passed a tipping point in their ability to use language. This improvement has been driven partially by fast computers and clever algorithms, like neural networks and word2vec. More fundamentally, it's been driven by data—our data. Talking machines don't give us a window on some new kind of linguistic genius. They only hold a mirror up to our own.

What will the future hold? It's impossible to know, of course, but a few likely trends stand out.

First, language models will become personalized; the machines around you will adapt to the way you speak, just as they adapt to your movie-watching preferences. As a result, they'll become much better at understanding you. Consider, for example, the historical trajectory of the iPhone. To use the iPhone 6, you had to teach it your thumbprint. To use the iPhone X, you had to teach it your face. It's not hard to imagine a future iPhone where you first have to read it a bedtime story, to teach it your voice.

Second, good policy and thoughtful regulations will be hugely important if we want to reap the benefits of these new NLP tools, without seeing them put to use in destructive ways. An algorithm that can write an episode of *Friends* seems cute, if a bit useless. That same algorithm will seem a lot more pernicious when someone can program it to flood the internet with fake news around election time. While we are ultimately optimistic, we are not policy experts, and we don't know the right answer to this kind of problem. We do know, however, that the problems themselves need to be part of the conversation. In the historical development of every new technology, from fire to gene splicing, there's always been a moment when "move fast and break stuff" ceased to be a morally tenable position for a grown-up. With computers and language, that moment has arrived.

But even as we recognize the potential downside, let's not forgot the upside, either. If you think that we have clever NLP algorithms and big data sets *today*, you ain't seen nothing yet. Think of the hundreds of millions of people out there dictating email to their phones, using

Google Translate, or talking to a bot on Facebook or WeChat. Each of these interactions leads to richer models and better performance, since these machines are just piggybacking on the trail of data we leave behind. As they improve—and there are a *lot* of improvements left to make—we expect these machines to become routine tools in every profession and in every part of life.

Call us heedless technological optimists if you want, but we think this is wonderful. Drudgery is not something we should wish upon anyone—and at least in the rich world, a lot of drudgery is the electronic kind, the kind that makes us sedentary and sick. Why should doctors spend hours every day doing data entry? Why should a blind person ever be forced to use a Braille keyboard? Why should a lawyer spend hundreds of person-hours sifting through millions of pages of documents? Why must office workers spend decades of their lives typing emails? Why should the EU spend hundreds of millions of euros per year translating everything into 23 official languages? Why must you rely on your half-wit thumbs to tell your phone what to do? And have you *seen* those computers they use at airline check-in desks?

We don't expect human beings to suck the dirt off a floor or filter spam from your inbox; we have vacuums and algorithms for that. Why should typing be any different?

Postscript

Grace Hopper may have been the first person to talk to a computer in English, but she certainly didn't stop there.

After her pioneering work on the UNIVAC, Hopper had a long career in private industry and in the Navy Reserves, before retiring in 1966 at age 60. Then, in 1967, she was unexpectedly recalled to active duty by the navy, where she served for another 19 years—long past the mandatory retirement age, by special approval of Congress. She helped wrestle the Department of Defense into updating its computing infrastructure, and she eventually became one of the first women in navy history to attain flag rank. At her promotion to commodore in 1983, her words as she shook the hand of President Ronald Reagan were "I'm older than you are." She finally retired for good in 1986, at the age of 79.

Hopper died in 1992, but her legacy lives on. Over the years she's had many things named after her, including a navy ship, a Cray supercomputer, and Grace Hopper College, at Yale University. She was posthumously honored with a Google Doodle in December of 2013, and with a Presidential Medal of Freedom in November of 2016. No doubt her great-grandfather the admiral would have been proud. Through her efforts to bring people and machines a bit closer together through language, Grace Hopper played an enormous role in inventing the modern world.

5

THE GENIUS AT THE ROYAL MINT

Real-time monitoring, from sports to policing to financial fraud: what Isaac Newton's worst mathematical mistake can teach you about the search for anomalies in massive data sets.

IF YOU'RE AN NFL fan, and you live outside a narrow strip of land from mid-Connecticut to Maine, then you probably view the New England Patriots—the most successful football team of the last 15 years—with a mix of peevishness and suspicion. First, there's all the winning, which is guaranteed to irritate the fans of all 31 other NFL teams. Then there's Bill Belichick, the Patriots' dour head coach, whose hoodie and scowl lend him an uncanny resemblance to the evil emperor from *Star Wars.* But even if you're not a fan of football, you can still be a fan of fair play—in which case the Patriots might still annoy you because of all their highly publicized cheating episodes, like for spying on other teams' practices or (allegedly) deflating footballs to gain an advantage in cold weather.

But could even the Patriots cheat at the pregame *coin toss*? Believe it or not, many people think so: for a stretch of 25 games spanning the 2014 and 2015 NFL seasons, the Patriots won 19 out of 25 coin tosses, for a questionably high winning percentage of 76%. As one TV commentator remarked when this latest "scandal" was brought to his attention: "This just proves that either God or the devil is a Patriots fan, and it sure can't be God."[1]

Before we invoke religion or the Force to explain this anomaly, let's consider the innocent explanation first: blind luck. Every time you call a coin toss, you have a 50% chance of winning. But there's also variability to account for. If you call a coin toss over and over again, you can easily get lucky stretches where you get it right more often than not, just by chance. Is it plausible that the Patriots just went on a lucky 25-game stretch?

The reasoning here is complicated by one fact: if the Patriots had gone on a suspicious 25-game coin-toss streak anytime since 2007, when their first cheating scandal broke, people would have noticed. Therefore, we can't just cherry-pick this particular 25-game stretch and ask how unusual it is in isolation. The right question to ask is: How unlikely is it for the Patriots to have won at least 19 coin tosses in *any* 25-game stretch over the last 11 seasons?

To answer this question, we used a computer to simulate coin tosses—more than 17 million of them. Specifically, we wrote a program to simulate a fair coin toss in all 176 regular-season Patriots games from the 2007 through the 2017 NFL seasons.* We repeated this simulation 100,000 times, checking each time whether the Patriots had a contiguous stretch of 25 games with at least 19 coin-toss wins. You can see 9 of these 100,000 simulations in Figure 5.1. Most of the time, the Patriots' 25-game coin-toss record hovers around 12 or 13 wins, which is what you'd expect. But there's a lot of variability. Sometimes the Patriots go on lucky streaks—as in simulations 3, 6, and 9, where they had 25-game stretches with at least 19 coin-toss wins. In sim 3, they even had a stretch with 22 wins.

Overall, the Patriots reached the 19-win threshold in 23% of our 100,000 simulations, which is hardly small enough to rule out luck as

---

* We also simulated the 24 games prior to game 1 of the 2007 season, so that the 25-game average was well-defined at the beginning of the 176-game stretch in question. This implies that the rolling 25-game winning percentage starting from that first game in 2007 actually went back to mid-2005.

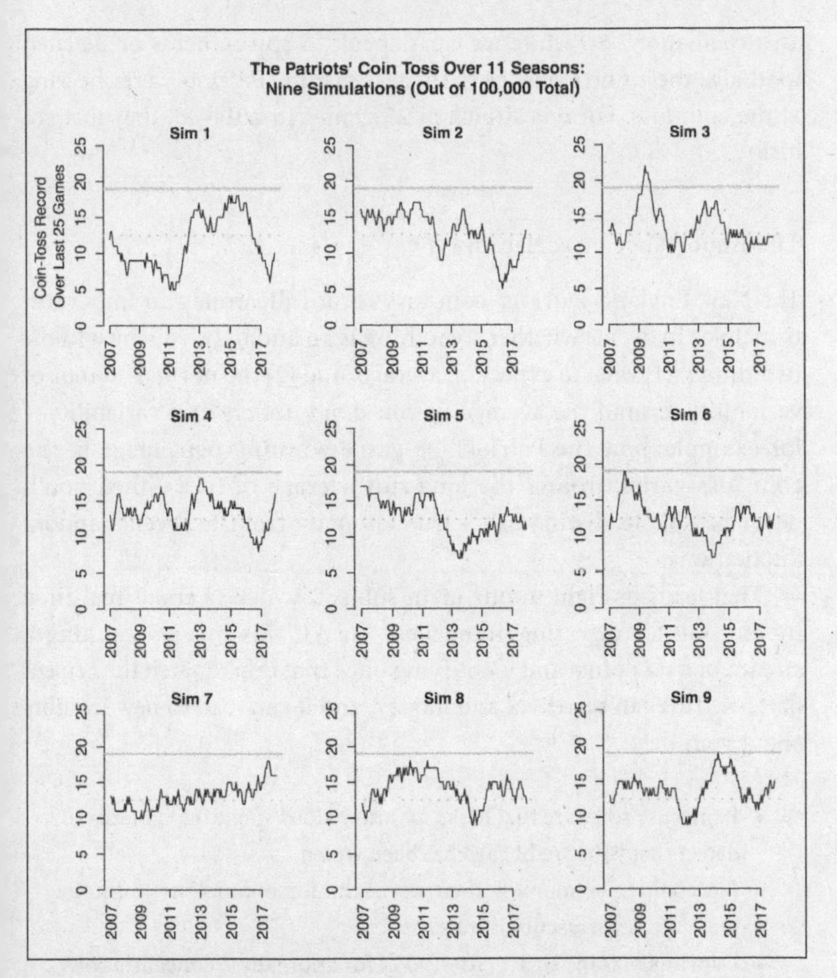

Figure 5.1. Each panel shows the Patriots' rolling 25-game coin-toss record over 11 simulated seasons, 2007–17. The vertical axis shows the number of wins over each contiguous stretch of 25 coin tosses. The horizontal gray bar is at 19 wins, while the dotted line is at the expected long-run average of 12.5 wins.

an explanation.[†] So while we can't speak to spy cameras or deflated footballs, there's no evidence to suggest that the Patriots were cheating at the coin toss. For one stretch of 25 games in 2014–15, they just got lucky.

The Importance of Variability

The New England Patriots' coin-toss record illustrates an important principle. To decide whether something is an anomaly, you must know two things: (1) what to expect on average, and (2) the normal bounds of variability around the average. If you don't understand variability—for example, how the Patriots' 25-game winning percentage in the coin toss varies around the long-run average of 50%—then you'll never be able to distinguish a true anomaly from innocent random fluctuations.

That leads us right to our main subject, which is about real-time monitoring for detecting anomalies.[‡] In AI, this means scanning a stream of data points and identifying ones that don't match the typical pattern. This can save lives and money, and it can lead to new insights about your data:

- Banks use software that looks for anomalous spending patterns to detect that your credit card has been stolen.
- Big companies monitor their networks for anomalous traffic to search for cybersecurity breaches.
- Data analysts in "smart cities" look for anomalous concentrations of crime in a given area to improve policing strategies.
- Investigators look for anomalies in medical-claims data to detect Medicare fraud.
- Sports teams monitor the data from their players' wearable gadgets, searching for anomalies that suggest a risk of injury.

---

† If you've taken a statistics class, you may recognize this number as the p-value ($p = 0.23$) under the null hypothesis of no cheating.

‡ Two synonyms for anomalies that you may have encountered are "signals in the noise" or "violations of the null hypothesis."

In all these applications of AI, and in thousands more, detecting anomalies is just as much about understanding the variability in your data as it is about understanding what's typical.

History's Oldest Anomaly-Detection System: A Lesson in What Not to Do

To illustrate the importance of this principle for AI, we'll start with an example of someone getting it very badly wrong—and not just anybody, but one of the greatest geniuses of all time. Specifically, we'll take you back to England, in 1696, when the longest-running anomaly-detection system in history was already in full swing. This system, called the "Trial of the Pyx," was designed to prevent fraud at the Royal Mint, where English money was manufactured. It is fascinating precisely because it failed: it missed anomalies left and right, for centuries on end, playing an important yet underappreciated role in an economic crisis that caused widespread suffering and anger.

And in 1696, the person at the center of it all was Isaac Newton.

Yes, *that* Isaac Newton—inventor of calculus, explainer of gravity, and the man immortalized in Alexander Pope's famous couplet: "Nature and nature's laws lay hid in night; God said 'Let Newton be' and all was light." In 1696, Newton was a 54-year-old scientific rock star, with a professorship at Cambridge guaranteed for life. He didn't have to teach, and he could work on whatever he wanted—physics, alchemy, apple juggling, anything. Yet in 1696 he gave up the life of a professor, moved to London, and accepted a sinecure offered by his powerful friends in government: warden of the Royal Mint.

Overall, Newton performed admirably in his new job—except on one crucial point, where he erred badly, by misunderstanding an important statistical principle that was staring him in the face for the better part of five years. Today, that principle sits at the heart of every real-time monitoring system powered by AI: in Silicon Valley, in smart cities, in the analytics office of every sports team, and in the fraud-prevention office of every bank. So if you want to understand any of these things, then you need to understand three major strands in the story of Newton at the Royal Mint.

1. A late-seventeenth-century crisis in the English economy, in which the Mint played a subtle but fundamental role.
2. The Great Recoinage of 1696, a drastic step in English monetary policy designed to stanch the crisis, and which Newton had to rescue from disaster.
3. The importance of statistical variability in detecting anomalies—the subject of the worst mathematical mistake that Newton ever made.

Isaac Newton's Second Career

Newton arrived at the Royal Mint in 1696 in the middle of a full-fledged currency crisis, one that threatened to bring England's economy grinding to a halt. To appreciate Newton's experience at the Mint, you have to understand the roots of that crisis.

The problem was this: by 1696, English money had been disappearing from circulation for at least three decades. At the time, England was on the silver standard, where the weight and silver content of coins determined their value. Due to the Nine Years' War, however, demand for silver on the Continent had skyrocketed, to the point where English coins were worth less as currency in England than as precious metal in Europe. In response, people in England did exactly what you'd expect. They took coins to France or the Netherlands, melted them down, exchanged the raw silver for gold, then sold that gold for more silver coins back in England—and presto, they were a bit richer than when they started. As all that silver leaked slowly across the Channel, England began, quite literally, to run out of currency.[2]

There was also a second reason why silver was disappearing, which massively amplified the effect of the first: coin clipping, the scourge of the English money supply throughout the 1600s. To clip a coin, you would find some protruding bit of silver along the edge. Then you'd just clip it off and file the damage smooth. You didn't get much from any one coin, but if you clipped enough of them, you could build up a healthy pile of silver. Clipping had been punishable by hanging since the reign of Elizabeth I. Yet that didn't seem to stop the clippers. In a parliamentary investigation in 1690, three goldsmiths each collected £100 of circulating coins; together these coins should have weighed 1,200 troy ounces, but they actually weighed 624. Thus clipping was obvious in the aggregate, but short of catching someone in the act, it was a nearly impossible crime to prove.[3]

Coin Clippers Loved Variability

Clipping was made far easier by the fact that before 1662, all English coins were struck by hand—that is, by a silversmith at an anvil, hammering a blob of molten silver into a disk. The key thing to emphasize about these hand-struck coins is their *variability,* in both shape and weight. This variability was essential to successful coin clipping. Variability in shape ensured that the coins had little bumps to clip in the first place. Variability in weight ensured that a clipper could always find a slightly heavy coin—and that once he'd clipped it, he could still spend it with an innocent shrug, as if it were merely a bit light to begin with.

In 1662, Parliament finally took notice of the clipping problem, by giving the Mint the funds it required to begin milling coins by machine. The goal was simple: put the clippers out of business by eliminating the variability in the shape and weight of coins.

This new machine-milling process is worth describing in detail, to give you a sense of the operation that Newton would later inherit. It started with molten silver at 1,000°C, bubbling away in giant cauldrons. The silver was poured into molds for ingots, and as these ingots cooled, they were flattened into a sheet by a giant pasta-roller contraption, powered by a team of four horses. A second cookie-cutter-like machine punched out disks from the sheet of silver. These disks were fed into a screw press, which made coin blanks. A fourth, very dangerous machine then stamped an image onto each coin's face. One man fed a blank into a small chamber in the middle of the assembly. Then four other men pulled ropes to turn a wheel through 180 degrees, causing an enormous press to stamp a deep, indelible image of the king's face onto the coin. Another half turn of the wheel brought the press back up—and during that interval, the first man had to flip the stamped coin out of the chamber and insert a new blank. The four men turning the wheel were driven to exhaustion after 15 minutes, and the man inserting coin blanks toiled in constant fear for his fingers.[4]

Finally came the edging machine, which inscribed two special features onto the edge. First, there was a pattern of ridges around the circumference; this is called a "milled edge," which you can still find on many modern coins, including an American quarter and a British £2 coin. Second, there was a Latin inscription: *Decus et Tutamen*, a phrase from Virgil's *Aeneid* meaning "an ornament and a shield." Just as the Latin suggested, it was a shield against coin clipping—for it would be hard to clip even a small bit off the coin without the damage being obvious.[§]

You might think that the introduction of these machine-milled coins would have helped solve England's currency problem, by eliminating variability in the money supply. In fact, after 1662, the problem got worse. It did so because the earlier hand-struck coins continued to circulate, and merchants continued to accept them at face value. By the late 1600s, England effectively had two parallel currencies. There were the pre-1662 hand-struck coins, so far debased that they could not be melted down at a profit. Then there were the machine-milled coins, which couldn't easily be clipped. So the machine-milled coins got melted down and smuggled to Europe, while only the hand-struck coins circulated.[5]

Economists refer to this phenomenon, where the bad money drives out the good, as Gresham's law. But this is a textbook case of economists naming obvious things after other economists. This "law" was in fact observed 17 centuries before Gresham, by Aristophanes, in his play *The Frogs*: "The full-bodied coins that are the pride of Athens are never used while the mean brass coins pass hand to hand." This is common sense. If you're paying for groceries and have the option of handing over a full-bodied Athenian coin or a mean brass one, you'll keep the full-bodied one for yourself. The shopkeeper will reason the same way when he gives you your change. Only the bad coins will circulate.

That's exactly what happened with English coins. As a result, by the

§ This inscription, *Decus et Tutamen*, remained on English coins into 2017, when it was sadly removed from the latest version of the £1 coin.

time Newton joined the Mint in 1696, the ordinary commercial life of England lay nearly in ruins. A bit of gallows humor went viral: that while taxes had been higher under King James, at least there had been money to pay them. Many people had no currency at all—and those who did would hoard it rather than spend it, reasoning correctly that it would be worth more tomorrow. As a result, wrote historian Charles Macaulay, "all trade, all industry were smitten as with a palsy. The evil was felt daily and hourly in almost every place and by almost every class." A contemporary witness explained to a friend that "no trade is managed but by trust. Our tenants can pay no rent. Our corn [merchants] can pay nothing for what they have had, and will trade no more, so that all is at a stand."[6] As Macaulay summarized, "all the misery which had been inflicted on the English nation in a quarter of a century by bad Kings, bad ministers, bad Parliaments and bad judges" paled in comparison to "the misery caused in a single year by bad crowns and bad shillings."[7]

The Trial of the Pyx

So far we've learned three important facts about the state of English money in 1696.

1. All those pre-1662 hand-struck silver coins were still a huge problem. They forced people into daily arguments over their worth, and their debasement drove the full-weight machine-milled coins out of circulation, and ultimately out of England.
2. These hand-struck coins had been debased by clipping.
3. The key to successful clipping was variability: unevenness in the shape and weight of coins that could be exploited by criminals, who ruthlessly clipped the slightly-too-heavy coins and passed them off as slightly-too-light coins.

Therefore, a crucial question for understanding the crisis in the English economy is: Why were the coins allowed to be so variable in the first place?

Some of this variability was inevitable, especially for coins struck by hand. English law had even evolved to recognize this, by setting a legal limit on acceptable variability, and by putting safeguards in place to make sure that these limits were respected. But the system failed—and to explain why this was happening, we come to the Trial of the Pyx.

The Trial of the Pyx* is an anomaly-detection system that's been operating continuously since the 1150s. Although the details have changed over time, its goal has always been more or less the same: to check whether the Mint has been cheating, inept, or both. The Mint officers, for example, could cheat by making the coins systematically too light, allowing them to pocket the leftover silver. Or they might simply be inept at quality control, making some coins too heavy and others too light. If this happened, then the accidentally-too-heavy ones could be melted down at a profit by some sharp-eyed merchant.

The Trial of the Pyx was designed to prevent such mischief. Out of every 60 pounds of silver struck by the Mint, a single coin was set aside. When several thousand coins had accumulated, typically every few years, they were tested for anomalies by a jury of silversmiths, to ensure that they met the legal standards of weight and metal purity.[8]

But remember the lesson of the New England Patriots' example: when checking for anomalies, you have to account for variability. Even if there had been no funny business, the coins couldn't be expected to weigh *exactly* what they were supposed to, because of the inevitable imperfections of the Mint's manufacturing process. Since at least 1345, English law had recognized this variability, by specifying allowable bounds for the weight of a coin. These bounds were called the "remedy," and they were set at about ±1% either way from the target weight.† If the coins fell outside these bounds, then the officers of the Mint would have to "make remedy" to the Crown for any shortfall—and they might be in for something much more ominous, since the Mint contract of

* The Trial owes its name to a room in Westminster Abbey, the Chamber of the Pyx. "Pyx" is an old Greek word for a box holding bread for Communion. Coins awaiting the Trial were stored in a box, and the box was stored in the chamber—hence Trial of the Pyx.

† We're rounding here to keep the numbers simple. The remedy was actually set at 48 grains per pound troy of silver, which is about 7 grams per kilogram, or 0.7% by weight. See Stigler, *Statistics on the Table,* chapter 23.

1280 placed them "at the prince's mercy . . . in life and members" if any wrongdoing came to light.[9]

Why Was the Trial of the Pyx So Ineffective?

From a data-science perspective, the Trial of the Pyx initially seems wonderful. It involves a well-defined sampling procedure with no obvious biases, along with a problem that any twenty-first-century statistics professor could assign for homework: the Crown officers calculated the average weight of a sample of coins, and they wanted to test whether that average was close enough to the target weight.

But what should count as "close enough"? That's where the Trial ran into trouble. The Crown officers assumed that the answer to this question was obvious: if the law says that a single coin must be within 1% of the target, then the average weight should also be within 1% of the target. This "obvious" answer, however, was badly wrong. It led to bounds for declaring an anomaly that were *way* too wide—and thus unintentionally favorable to coin clippers.

To understand the error here, imagine that you have a sample of 2,500 shillings in front of you. Suppose that each shilling is supposed to weigh 100 grams—the "expected value"—with an allowable margin of error of ±1 gram. Your job is to see whether these coins fall within ±1 gram of their expected value. The obvious approach is to weigh all 2,500 coins individually. But can you imagine how tedious that would be? It's the 1600s, after all; you've got lots of other ways to spend your time, like playing the lute or attending a public execution. So you decide to save all that work by calculating the *average* weight of the coins: you weigh them all at once on the same scale, and you divide the result by 2,500. This seems to offer a pretty good snapshot of the Mint's work. If the average is close to 100 grams, then most of the individual coins must also be pretty close to 100. If the average is far from 100 grams, then at least some of the individual coins must be far from 100, too.

This averaging procedure—which is almost exactly how the Trial of the Pyx actually worked—will be certain to detect the most obvious anomalies. Suppose, for example, that the average weight was only 50 grams, versus the target of 100. This is pants-on-fire stuff: you should advise the Mint officers that they need to hire a good lawyer with all the silver they've stolen, at least if they value their lives and members. Now, what about a case where it isn't so obvious—say, if the average weight was 99.5 grams? This might initially seem like evidence of cheating, too, just like the Patriots' coin-toss record of 19 wins in 25 games. But remember, some "anomalies" are just down to luck. How can we decide whether an average of 99.5 grams is actually suspicious?

Here's the key question. If the allowable bounds for a *single* coin are ±1 gram from the target, then what should the bounds be for the average weight of *many* coins? There's a Goldilocks principle here. Suppose you make the bounds very narrow, allowing the coins to pass the Trial only if their average weight falls within ±0.0001 grams of the 100-gram target. Then your anomaly-detection system will run too hot: it will detect anomalies everywhere, like those early car alarms from the 1980s that would go off if you touched the car with a feather. On the other hand, suppose you make the bounds very wide, like ±10 grams—about the weight of a lemon wedge. Then your system will run too cold, causing you to miss real anomalies. The million-shilling question is: If ±0.0001 is too hot, and ±10 is too cold, then what's just right?

SIDEBAR: THE SQUARE-ROOT RULE, A.K.A. DE MOIVRE'S EQUATION

There's a very important equation in statistics, called the square-root rule, that says *exactly* how tight the bounds for declaring an anomaly should have been in the Trial of the Pyx. This equation was discovered by Abraham de Moivre, a Swiss mathematician, in 1718. In our opinion, it represents one of the most underrated triumphs of human reasoning in history. Most people, for example, have heard of Einstein's equation: $e = mc^2$. De Moivre's equation is just as profound—it represents an equally universally truth, and is equally useful for making accurate scientific predictions. Yet very few people outside

(continued on next page)

statistics and AI know it. This is a shame, given how central it is in the age of learning machines.

De Moivre's equation establishes an inverse relationship between the variability of a sample average and the square root of the sample size. It goes like this:

$$\text{Variability of an Average} = \frac{\text{Variability of a Single Measurement}}{\sqrt{\text{Sample Size}}} = \frac{\sigma}{\sqrt{N}}$$

Data scientists usually use the Greek letter σ (sigma) to denote the variability of a single measurement, and the letter *N* to denote the sample size. So that's why we've expressed this equation both in words and a little more compactly in symbols, as $\sigma/\sqrt{N}$. Data scientists refer to the "variability of an average" using a slightly more technical-sounding term: the "standard error of the mean."

Let's try an example with some actual numbers attached. Suppose you're weighing 2,500 shillings ($N = 2{,}500$). The law allows each coin to vary by as much as ±1 gram from the target weight of 100, on average. Thus if the Mint's quality control is up to snuff, then $\sigma = 1$. The square-root rule says that, if this is true, then the average weight of the 2,500 coins should fall within $1/\sqrt{2500} = 0.02$ of the expected value of 100. Thus the bounds should be 100 ± 0.02. Anything outside those bounds suggests one of two possible anomalies: either a "bias," meaning that the average weight of the coins isn't actually 100, or "over-dispersion," meaning that the variability of a single measurement is actually higher than 1.

As we've said, the people running the Trial of the Pyx believed that if the allowable bounds for a single coin were ±1 gram, then the bounds for the average weight of many coins should also be ±1 gram. But according to modern statistics, this was a bad mistake. Instead, the bounds should depend on how many coins are in your sample: the bigger the sample, the tighter the bounds. This is a consequence of a very important equation called the "square-root rule," also known as "de Moivre's equation." This rule says that the variability of a sample average gets smaller as the square root of the sample size gets bigger. The math here is a bit complicated, but the intuition is simple. In a small sample, a single coin that's too light can bring the average down a lot. But in a large sample, a light coin is very likely to be balanced out by a heavy one, so the average should be closer to the target. So if you average thousands of measurements and the result isn't very close to what you expect, then something is fishy.‡ This is the same math that casinos use when they decide whether to send some burly dudes to visit the MIT kids at the blackjack table.

To show you how important the square-root rule is for reasoning about anomalies, let's see how it plays out, by comparing two sets of bounds side by side.

Sample size	Bounds used in the Trial of the Pyx	Correct bounds based on modern statistics
1	100 ± 1.00	100 ± 1.00
100	100 ± 1.00	100 ± 0.10
2,500	100 ± 1.00	100 ± 0.02
10,000	100 ± 1.00	100 ± 0.01

The Trial of the Pyx was using bounds that were *far* too wide. This mistake would have allowed for two possible kinds of undetected anomalies, both of them bad for England.

First—and this probably didn't happen, but it's the possibility that arouses the interest of pretty much every data scientist who ever hears of the Trial of the Pyx—the Mint officers could have been skimming silver. To see how this might work, suppose the Mint was capable of striking coins to the legal standard of variability (±1% by weight), but that the officers sneakily aimed for a target of 99.5 grams per shilling, rather than 100. (This kind of anomaly is called a "bias" of 0.5 grams.)

‡ If you want to see the math behind the square-root rule, see the technical sidebar on page 156. But following the math here isn't necessary to get the main lesson of this chapter.

Let's also suppose that the Trial of the Pyx were trying to detect this fraud by weighing a sample of 2,500 coins. If you churn through the math of the square-root rule, you find that the average weight of these 2,500 coins will likely fall between 99.48 and 99.52 grams. This is far outside the statistically correct bounds of 100 ± 0.02. Yet the Trial of the Pyx would have failed to raise an alarm, since the jury would accept any average between 99 and 101 grams. Enterprising Mint officers could, in theory, have pinched 0.5% of all the silver in England without getting caught. But only someone who knew the square-root rule could have exploited the Trial of the Pyx in such a clever way, and there is no evidence to suggest that such a blockbuster fraud ever took place.

There *is* evidence, however, that a second, subtler kind of anomaly really did happen: that the Mint was honest but sloppy, coining money of highly uneven weight and thus giving coin clippers the gift of extra variability. Suppose, for example, that the Mint did indeed aim for the target weight of 100 grams on average. Now imagine that as a result of poor quality control, the coins were 10 times more variable in weight than the law allowed: 100 ± 10 grams, rather than 100 ± 1. This kind of anomaly is called "overdispersion." It isn't necessarily evidence of fraud, just carelessness. But the Trial of the Pyx would still have been unable to detect it. If the coins individually fall within the bounds of 100 ± 10, then the square-root rule implies that the *average* weight of 2,500 such coins would almost surely fall between 99.8 and 100.2. Again, this is quite likely to fall outside the correct bounds of 100 ± 0.02, and would therefore be detected by a modern quality-control procedure. But the Trial would accept anything between 99 and 101.

In an enormous boon to coin clippers, this kind of overdispersion anomaly almost certainly persisted for decades, if not centuries. We base this conclusion on two facts. First, Isaac Newton explicitly remarked on the poor manufacturing standards he discovered upon arriving at the Mint. He even paid special attention to the coins' variability: "When I first came to the Mint, *and for many years before*," he wrote, "the money was coined unequally, pieces being two or three grains too heavy and others as much too light."[10] Newton also wrote that the heavy coins "were called 'Come Again Guineas' because they were culled and brought back to the Mint to be recoined" at a profit for someone else.

He estimated that the fraction of "come-again guineas" was as high as *one coin in four*—a smoking gun of excess variability. This was hardly a new issue at the Mint. At the Trial of the Pyx in 1534, for example, the jury had specifically remarked that "the coins were very uneven, so that it was profitable to pick out heavy pieces."[11]

Second, by Newton's time there had been two historical occasions when coins had failed the Trial of the Pyx.[12] Two failures may not sound like a lot, but if the individual coins had actually met the legal variability standard of ±1%, then in light of the Trial's overly wide bounds, even *one* failure would have been far more improbable than your winning the lottery. In light of Newton's comments about poor quality control, the simplest explanation for this high failure rate is that the coins were far more variable than the law allowed.

Newton at the Mint

The officers of the Mint, it turned out, were really bad variability police. They let the silversmiths get away with decades of slapdash work, producing coins that were much more variable than the legal standard of ±1% by weight. Yet the Trial of the Pyx *never held them to account*, thus giving coin clippers an enormously powerful ally: the laws of probability.

The math behind this dreadful error, however, was not something that any of the Mint officers over the centuries could have appreciated—with the one very notable exception of Isaac Newton.

You must pity poor Newton upon his arrival at the Mint. His new job wasn't quite the plum he'd been promised. He'd been told that it paid £600 a year, but this had been a deliberate exaggeration by the chancellor of the exchequer, for it paid only £400. He'd been told that his new colleagues were a team of sharp-minded professionals, when in reality they were an incompetent rabble; the deputy master in Norwich ended up in jail with his property confiscated, while the deputy controller was soon dismissed from his post and subsequently reappointed as His Majesty's Ambassador to the Pirates of Madagascar.[13] Finally, Newton had also been told that he wouldn't have to work very hard in his new job—and this was the biggest lie of all. He arrived at the Mint during the Great Recoinage of 1696: Parliament's drastic, all-hands-on-deck solution to the coin-clipping problem, in which all the millions of hand-struck coins in England were called in to the Mint, melted down, and recast by machine.

The Great Recoinage was in full swing when Newton arrived, and by all accounts it was headed for disaster, owing to bad leadership. Yet Newton, to his credit, did not treat his new job like the sinecure it was. Instead, he leapt into action. He took on extra work when his colleagues failed him. He mastered every detail of the Mint's complex accounting system. He suggested improvements based on his knowledge of metallurgy, which he'd honed during his many years spent in pursuit of alchemy. This knowledge never helped turn lead into gold, but it certainly did help Newton turn silver bullion into coins.[14]

Then there was the pace of the Great Recoinage. Remember that hellish mechanized coining operation, with the two-story pasta roller and the machine that ate men's fingers? The workers had assumed that 3 or 4 coins per minute was an acceptable pace, but this was clearly too slow to complete the Great Recoinage in time to avoid disaster. So Newton personally undertook a detailed time-and-motion study of the workers on the line, and his changes ultimately spurred them to a pace of 50 coins per minute. This clip was maintained from 4:00 A.M. to midnight, seven days a week, for almost two years.[15]

By 1701, the Great Recoinage was done, and there were no more hand-struck coins in England.[16] Newton had been promoted to the much more prestigious job of master of the Mint, occasioning a Trial of the Pyx. The jury convened, the coins passed the test, everyone had a big dinner on the new master's dime, and that was that. Newton complained bitterly about the cost of the dinner: £2 for each jury member, or over £200 a head in today's money.[17]

Thus Newton's Trial of the Pyx is remarkable precisely because it happened with a whimper rather than a bang. Here was Isaac Newton, the best variability policeman in the history of the Royal Mint. He'd spent five years contemplating every detail of the coining process. He had specifically remarked that the coins were too variable to meet the legal standard. He'd recognized that excess variability had been a problem at the Mint for a long time, and he had become obsessed with reducing that variability. Finally, he was the greatest mathematician in the world, facing a public trial with serious consequences, in which the variability of the coins was *exactly* the issue in question.

If ever there were a case of the right person in the right place at the right time to make a fundamental discovery in statistics, this was it. Yet nothing happened. Newton betrayed no recognition that there even was a problem in need of a solution—and the Trial of the Pyx kept making the same mistake for another century. Why didn't Newton discover the square-root rule? This is a mystery. It is hard to believe that a simple empirical question failed to occur to Newton: If the individual coins were drastically more variable than the legal standard, as he'd explicitly pronounced, then why had they passed so many Trials of the Pyx, for hundreds of years on end?

This is especially puzzling in light of Newton's lifelong appetite for math problems, which remained unquenchable even during the frenzy of the Great Recoinage. On one afternoon in 1696, for example, he came home from the Mint at 4:00 P.M. and sat down to work on a famously hard problem, called the "brachistochrone curve," that had been posed by Newton's snarkiest math enemy, Johann Bernoulli.[§] Newton was very tired from his work at the Mint that day—but as he wrote in his diary, he was even more tired of being "dunned and teased . . . about mathematical things." So on that day, he skipped dinner and refused to rest until 4:00 the next morning, by which point he had solved the problem and shown Bernoulli who was boss. This kind of thing was typical for Newton, even in his so-called retirement.[18]

So it wasn't for lack of opportunity or genius or tenacity, or the motivation provided by a real math problem right in front of him, that Newton failed to see his error. These are all things that we point to as the essential ingredients in a scientific breakthrough. The case of Isaac Newton at the Royal Mint had all these ingredients, but no breakthrough. The irony is that, compared with the brachistochrone curve, the math behind the square-root rule would have been a piece of cake for Newton—if he'd just thought to ask the right question in the first

§ Bernoulli thought Newton's theory of gravity was nonsense, and he was also buddies with Leibniz, who had a priority dispute with Newton over calculus.

place. But he didn't, and for that matter neither did anyone else for a very long time. Probability and statistics wouldn't even be formalized as a discipline for nearly another century, and it fell to two great mathematicians of a later age—Gauss and Laplace—to articulate the full implications of the square-root rule.

Anomaly Detection in the Age of AI

Newton's time at the Royal Mint is a fascinating and surprisingly little-known historical episode, and also one that has major implications for artificial intelligence. Averaging lots of measurements together is the most important idea in the history of data science. A huge number of applications of AI depend upon this idea, from fraud prevention to smart policing, and they all work along the same basic lines as the Trial of the Pyx.

- Data collection: Several measurements are taken of some underlying process.
- Averaging: Those measurements are averaged together to provide a "numerical snapshot" of the process.*
- Decision-making: Is the average "close enough" to what we expect, or is it outside the bounds of normal variability?

There are three big differences from Newton's day. The decision to flag an anomaly is usually made by a machine rather than a person. This decision happens on a time scale of a few milliseconds rather than a few years. Finally, unlike the people running the Trial of the Pyx, these machines get the math right.

These AI systems have become ubiquitous. Formula 1 teams monitor the data streams from hundreds of sensors on their cars to look for anomalies—engine temperature, tire wear, aerodynamics, anything

* Many of these systems don't rely on an average per se but on some other numerical snapshot of the data. A simple example would be a median, while complicated examples have names like "principal component scores" or "Kolmogorov–Smirnov statistics." This is an unimportant detail; the need to understand variability remains the same, whether you're using an average or some other fancier numerical snapshot.

that might affect race tactics. Credit card companies scrutinize every transaction you make, looking for anomalies that suggest fraud. Police officers in big cities carry radiation sensors programmed to look for anomalies that might indicate a dirty bomb left by a terrorist. Facebook and Google, warehouses and grocery stores, airlines and oil rigs, senators and stock traders, the Cleveland Cavaliers and the weekend warriors . . . they all take measurements and average them together, algorithmically searching for anomalies in massive data sets.

While the speed and scale of these systems have changed a lot in three centuries, the fundamental principle hasn't: to detect an anomaly, you have to understand variability.

Smart Cities: Big *N*, Big *D*

Just ask the folks who work at the Mayor's Office of Data Analytics, or MODA, in New York City. MODA was created in 2013 by then-mayor Michael Bloomberg to analyze the vast troves of municipal data collected by the city government—everything from 911 calls to building-inspection forms to horticultural reports on the city's 5.2 million trees.

The richness and scale of MODA's various data sources reveal an important fact about the intersection of big data with artificial intelligence. Big data sets don't get that way merely because of a "big *N*": the number of data points they have. They also involve a "big *D*": the number of details recorded about each data point. For example, in the case of a data set about apartments in New York, the details might be size, location, and amenities; in a data set about surgery patients, the details might involve a set of health indicators. "Big *N*" means lots of data points—lots of apartments, lots of surgery patients, and so on. "Big *D*" means lots of detail.

We can think of a "big *N*, big *D*" data set as a collection of many smaller subsets that collectively exhibit a dizzying breadth (big *N*) and a hyperspecific combination of details (big *D*). As a result, using artificial intelligence on this kind of data set is rarely about searching for one anomaly in an ocean of data; rather, it's about searching for thousands of possible anomalies in millions of different ponds. The bigger and richer the data set, the more ponds there are, and the more detailed the anomalies you're able to find.

For example, New York City has only about 200 building inspectors to investigate more than 20,000 complaints each year regarding illegal apartment conversions, like where a landlord transforms an industrial space into a residence or chops up an already-small apartment into tiny subunits.[19] These inspectors must be smart about how they allocate their resources, so they turned to the Mayor's Office of Data Analytics to help them find which features of a property are most likely to yield a "hit": an inspection that finds an illegal conversion.

To see how this works, imagine that the inspectors' historical hit rate is 10% across all apartments. Now consider the following subsets of apartments that, based on previous inspections, seem to have elevated hit rates.

- Subset A: fifth-floor walk-ups below Fourteenth Street built before 1940, with ground-floor retail. Hit rate = 2 out of 10 (20%).
- Subset B: new-construction two-bedroom condos in Queens. Hit rate = 17 out of 100 (17%).
- Subset C: disused garment factories with more than five new restaurant-permit applications within a five-block radius. Hit rate = 2 out of 5 (40%).

All of these subgroups have hit rates that exceed 10%, but only one is an anomaly—a hit rate that's hard to explain as a result of random chance. Which is it? Before we get to the answer, let's emphasize the key insight: checking any one of these subsets for an elevated hit rate is like conducting a single Trial of the Pyx in miniature. The goal is to test for an anomaly, or a difference from the overall hit rate of 10% that is too large to be explained by chance. (Hint: Pay attention to the sample sizes.)

You might assume that the anomaly is the one with the highest hit rate: subset C, at 40%. But according to the square-root rule, it's actually the one with the *lowest* hit rate: subset B, at 17%. The reason is that subset B also has the largest sample size (100), which means we're pretty sure that the elevated hit rate is real. The hit rates in subsets A and C, on the other hand, could be elevated because of sampling variability—that is, which particular apartments you happened to have inspected already. This recalls the lesson of the Patriots' coin-toss example: variability is really important for detecting anomalies, and small samples can be highly variable.†

Of course, in the real word there are many more than three subsets of apartments. That's why we turn to AI, which can search for anomalies across thousands or millions of possible combinations of building features, including ones that a human may never have imagined to be important. The design of algorithms capable of doing this accurately and efficiently continues to be a major area of research. (We'll spare you the gory math details.)

When the folks at MODA started applying these algorithms to correlate building-inspection reports with New York's vast collection of other municipal data sources, they produced astounding results. The inspectors increased their hit rate by a factor of *five*, and they found two factors that correlated especially strongly with illegal apartment conversions: sudden spikes in utility bills and increased reports of sanitation issues. Another team of inspectors used the same techniques to search for stores selling alcohol and tobacco illegally, and they also improved their hit rate: from 30% to 82%. Yet a third team managed to put a dent in opioid abuse, using Medicaid claims data to identify a small set of pharmacies—about 1% overall—that accounted for 60% of the city's Oxycodone prescriptions.[20]

And that's just the inspectors. Imagine the potential for similar improvements as other city agencies start to collect and monitor new kinds of data—from the police to the pothole fillers, from the parks department to the fire brigade. Imagine, for example, how many lives might be saved if you could identify not just where and when but *why* people tend to get hit by cars frequently. Then you'll begin to understand why city governments across the world are raising a toast to the power of AI.

---

† These other subsets might turn out to be anomalies; we just need more data to be sure. In practice, there's a trade-off between exploring low-data subsets versus getting "easy wins" by inspecting subsets with clear anomalies.

Sniffing for Gamma Rays and Gas Leaks

One of the people they toast might soon be Alex Reinhart, a PhD student in statistics at Carnegie Mellon University who is working on a new anomaly-detection system that may someday help law enforcement officers sniff out one of the most odious of all terrorist threats: a dirty bomb.

A dirty bomb is a cruel and deadly weapon that uses a conventional explosive to disperse radioactive material through the air. The initial explosion would destroy a small area and would poison a much larger one—perhaps dozens of city blocks. But the potential good news for law enforcement is that any radioactive isotope emits gamma rays at predictable energies, in a way related to that isotope's atomic structure. The anomalous gamma rays emitted by a dirty bomb could therefore, in principle, be detected by a radiation "sniffer" before the bomb goes off.

Yet there are three catches—three sources of variability that make it hard to flag an anomaly. First, you can't raise an alarm anytime radiation is detected, because background radiation is everywhere. Most building materials, like bricks and stone, have tiny amounts of radioactive uranium and thorium. Bananas and garden soil have traces of radioactive potassium. Not to mention that gamma rays arrive all the time from outer space. Researchers use the term "NORM," for "naturally occurring radioactive materials," to refer to these benign sources of gamma rays. They're harmless, but they mean you can't just flag an anomaly if you detect radiation.

Second, this background radiation varies from place to place, especially in a big city. If you walk across the street or around a corner on your daily bomb patrol—a sad necessity for law enforcement in many cities—you'll find yourself near different buildings made of different materials, each with a slightly different NORM profile.

Finally, radiation is statistically noisy, for reasons ultimately having to do with quantum mechanics. A radioactive isotope emits a random number of gamma rays at random energies over any given time period. Therefore, you can never be sure whether any particular gamma ray came from the background or an anomaly.

The upshot is that looking for radiation anomalies is a surprisingly tricky data-science problem. You must compare the observed radiation, which is variable, to the normal background radiation, which is also variable, and which changes from place to place. To do this, you need a detailed map of the background radiation for the entire city, together with a good algorithm for detecting small anomalies in noisy data.

Currently, the best way forward involves human intelligence—basically, hiring someone with a PhD in nuclear physics to monitor the readings in real time. But this is hardly a scalable solution for the kind of antiterrorist operations you'll find in London, New York, or Paris, where you'd need a small army of people with this high-level skill.

Reinhart and his collaborators propose to use artificial intelligence instead. They envision an officer equipped with a small gamma-ray sniffer, hooked up to a smartphone with a GPS sensor. Every two seconds, the smartphone uploads the reading from the gamma-ray sniffer, together with the officer's GPS coordinates, to a central server. The server queries a geospatial database of the city's background radiation, compiled over many months using cheap mobile sensors. The current reading is compared with the typical background at the officer's location, using the square-root rule to determine the bounds for declaring an anomaly. If those bounds are exceeded, the AI system alerts the officer to investigate.

The applications of this geospatial-aware sniffing technology don't stop merely at bomb detection. One of Reinhart's research mentors, Dr. Alex Athey, points out that every major city has a vast infrastructure of natural gas pipes, all of them prone to potentially dangerous leaks. For example, New York City has over 6,000 miles of gas pipes beneath its streets, and it discovered 9,906 leaks in 2012 alone.[21] In March of 2014, one such leak caused an explosion in East Harlem that killed eight people.

While a city could install a new network of "smart pipes" capable of raising an alarm if they spring a leak, this would be both disruptive and very expensive. Athey's proposed solution is much cheaper. Imagine placing methane sensors on ordinary municipal vehicles, like garbage trucks, city buses, or ambulances. Over time, these vehicles would collectively traverse much of the city, building up a background map of "normal" low-grade methane levels. If a gas pipeline somewhere sprang a leak, those same mobile sensors would likely register the anomaly much faster than the gas company ever could—and much more cheaply than retrofitting thousands of miles of pipes buried several feet underground.

Fraud Detection Today

City inspectors and police officers aren't the only ones looking for lawbreakers by tracking the anomalies they leave in massive data sets. In that effort, they are also joined by the world's largest banks, who are increasingly turning to artificial intelligence to help them ward off the curse of the modern digital economy: fraud.

Fraud might be the world's second-oldest profession. The ancient Greeks believed in Apate, the goddess of deceit and one of the evil spirits found in Pandora's Box. The Egyptians employed an entire class of scribes to monitor transactions involving the pharaoh's grain inventory. And someone must have done something 3,000 years ago to provoke King Solomon, who said in Proverbs 11, "A false balance is abomination to the Lord: but a just weight is his delight."

Until quite recently, the battle against fraud happened in person, using human intelligence. In 1685, your silver coins would be inspected and weighed. In 1885, your promissory note was as good as your reputation. In 1985, your personal check would be accepted only if it matched the details on your driver's license. Today, however, it's a chip-and-PIN world out there, and this kind of face-to-face vigilance is no longer feasible. American banks, for example, processed $178 *trillion* worth of noncash transactions in 2015. That figure includes 70 billion individual swipes of a debit card, 34 billion swipes of a credit card, and 24 billion individual bank transfers.[22] Alas, it also includes billions of dollars in fraudulent transactions—most of them at ordinary retail merchants, who pass on those costs to you. For every dollar you spend at the grocery store, 1.3 cents go to e-crooks. They're the coin clippers of the modern world.

Luckily, data scientists are hard at work on AI systems capable of fighting back. The key, as it is with all forms of anomaly detection, is measuring variability. Think about your own spending habits, which vary in predictable ways from day to day and week to week. That variability forms the statistical baseline against which fraud can be detected.

For years, every major bank has analyzed its credit and debit card transactions in real time to look for fraud, which is why your card sometimes gets declined. But most of these legacy systems rely on simple fixed rules, like the size and location of the transaction. This ignores a lot of important person-to-person variability. For a teacher, a string of card payments in three different countries in the middle of a school week might be a clear signal of fraud. For a traveling sales rep, that same pattern might be normal—and the difference between these two customers should be obvious from their transaction histories.

You might think that the credit card companies would have been mining your transaction history to learn about those differences for a long time now. Indeed they have been—but only sort of, and mainly for grubby marketing purposes. Unfortunately, it has proven much more difficult to leverage all that data in a real-time payment system that must accept or decline your card within a tenth of a second.

The reason is simple: the colossal engineering challenge posed by working with data sets on such a large scale. Credit card companies generate petabytes of transaction data, and one petabyte is about 220,000 DVDs. Until recently, no end-to-end AI systems were fast enough to leverage all of that data in support of sophisticated real-time fraud detection. They all suffered from some decisive weakness—whether the performance of the fraud-detection algorithm itself, the speed of the network, or the surprisingly slow process of reading all those countless trillions of 1s and 0s off a physical disk.

As a result, banks faced a compromise. If they wanted to analyze 100 billion transactions, each in milliseconds, they were stuck with relatively basic "small *D*" anomaly-detection rules based on timing or location or dollar value. And if they wanted to exploit the fantastic level of detail found in each customer's unique transaction history, they would need months, not milliseconds, to look for anomalies. They could choose big *N* or big *D*, but not both.

PayPal, however, is just one of the many payment-system companies to have finally solved this problem, with the help of modern algorithms and a modern supercomputing infrastructure. Its fraud-detection system uses deep learning to compare every transaction with your own past behavior, as well as the behavior of other users similar to you. On the basis of that comparison, which uses thousands of possible features, the system produces a fraud-probability score that can be used to accept or decline the transaction—all in a fraction of a second.

With this new system in place, PayPal now has a much better appreciation of the normal bounds of variability in its data, down to the level of an individual user. Its investment in AI has paid off handsomely: its fraud rate dropped to 0.32% of revenue in 2016, less than a quarter of the industry-wide average.[23] Other payment-system companies, like Alipay in China or Stripe in the United States, have invested in similar technologies. And these systems keep improving, since they learn a little bit more about fraud with every new data point.

King Solomon and Isaac Newton would both be proud.

Moneyball for the Digital Age

If you're a sports fan, you've probably heard of "Moneyball," author Michael Lewis's term for a particular data-driven approach to building and coaching a sports team. In the late 1990s, the Oakland A's figured out that traditional baseball scouts weren't actually very effective at assessing what made a good player. A lot of what these scouts attributed to skill was really luck, and vice versa; they were systematically confusing signal with noise. On the advice of their scouts, most baseball teams paid millions of dollars for players who didn't actually help to win many games, or whose past successes weren't reproducible. Meanwhile, many other players flew under the radar, despite helping their teams in ways that were important and reproducible but not obvious. This inefficiency created an opportunity for the first team to figure out a better way.

Oakland's innovations were threefold. They used data to determine which player characteristics and habits actually won games. Then they used those findings to look for market anomalies—winning characteristics and habits that were systematically undervalued by other teams. Then they hired players with those characteristics and coached them to have those habits. As a result, they were able to compete—and win—against teams like the Red Sox and Yankees, who could afford to spend three times as much as Oakland on players.

If you fast-forward 25 years, these innovations have changed every major sport in the world. Today, though, there's one big difference. In the 1990s, Moneyball was something you could play with a spreadsheet and a smart intern. Now you need a cloud-based supercomputer and a dedicated team of data scientists—all because of the massive new data sets that professional sports teams began collecting once they realized the edge it would give them.

Formula 1

The poster child for this revolution is Formula 1, the most popular auto-racing competition in the world. At a Formula 1 race, the data flows faster than the champagne in the luxury boxes. Every aspect of a car's performance is monitored, in real time and in microscopic detail. A Formula 1 car generates several gigabytes of data per lap, roughly the amount needed to stream 30 hours of songs or download 6,000 e-books. This data is beamed wirelessly back to the pit crew, who use sophisticated algorithms to look for anomalies that might affect race tactics: engine power, brake temperature, fuel consumption, tire wear, lateral g-forces, downward force on the rear wing, and hundreds more variables. Teams no longer have to wait until a part fails unexpectedly, ruining their race. Now they can predict those failures before they happen.

In fact, the data mining doesn't stop at the track. Formula 1 involves a costly high-tech arms race among the teams, and to limit this, the rules cap the number of trackside personnel that each team may have on race day. Without this, the big teams would spend the small ones into oblivion; after all, this is a sport where teams pay $100 million a year for engines and employ three people *per tire* for pit stops. The richest teams have decided, however, that they need even more number-crunching horsepower, so they turn to off-track engineers. Red Bull Racing, for example, recently partnered with AT&T to build a global network for transmitting race data from any Formula 1 track in the world to its team headquarters in Milton Keynes, England. There, a second team of data scientists monitors the Red Bull car in real time. Or nearly real time, that is: the limiting performance factor of the system is the speed of light, which can travel around the world only 7.5 times in a second. That kind of investment should give you a good sense of how much real-time anomaly detection means to the engineers.

Formula 1 teams have gotten so good at real-time monitoring that some of the better ones have started to sell their services to other big companies. The McLaren team, for example, recently spun out its data-analytics team into a separate company, called McLaren Applied Technologies, which immediately signed a deal with the consulting firm KPMG. Among other projects, it's now helping clients in the oil industry monitor real-time sensor data from drilling rigs, looking for anomalies that might suggest trouble.

Beyond the Racetrack

These innovations have bled over into other sports. In 2016, for example, the Brooklyn Nets signed a sponsorship deal with a company called Infor, barely known to those outside enterprise-software circles. Infor builds software for big-data analytics—including for Ferrari, a Formula 1 team—and while it paid millions of dollars for the right to show its logo on the Nets' jerseys, it also brought much more to the bargaining table than simply an open checkbook.

Brett Yormark, the Nets' CEO, explained that in selling the real estate on his team's jersey, he wanted to identify a strategic partner "that was substantive enough to help us with performance both on and off the court." The deal he signed with Infor is emblematic of the NBA's new era of Moneyball, in which some of the league's biggest stars will wear the logo of a big-data company on their jerseys.[24]

In the NBA, much of this revolution has been fueled by new data sources, like motion trackers on every player and cameras that cover every angle of the court. But it's also been fueled by a widespread change in team-building philosophy, and by a major investment in analytical talent. The Sacramento Kings, for example, recently hired Luke Bornn, formerly an assistant professor of statistics at Harvard, to mine all of that video and player-tracking data. As Bornn told NBC Sports in an interview:

> There is a lot of what happens on the court that really is not picked up by the box score. A lot of players that make big contributions make it in ways that don't appear. It's not an assist, it's not a rebound, it's not a block.[25]

Rather, it's something else—something previously unrecognized by coaches, but hidden in plain sight in the data and waiting to be discovered. Bornn is convinced that using artificial intelligence to mine all that data for interesting anomalies will help the Kings find undervalued players and coach them in innovative ways. He and a group of coauthors, for example, recently published a paper on advanced measures of defensive skill in basketball. Using data captured by cameras mounted in the rafters of all NBA arenas, they were able to answer two simple questions that had never played a role in basketball statistics before: Who was guarding whom at every moment, and how well did specific defenders do against specific opponents?

Bornn and colleagues found that shot *selection* (when and where a player shoots) and shot *efficiency* (whether the shot is made) are two distinct components of defensive skill in basketball. These skills also have a clear spatial structure: they depend on where a defender is positioned on the court. Near the basket, for example, center Dwight Howard of the Charlotte Hornets is better than average at reducing shot frequency, but worse than average at reducing shot efficiency—and he's below average at both when he's far away from the basket. These findings allowed Bornn and colleagues to predict the results of specific defensive matchups. For example, their model inferred that LeBron James should be expected to score fewer points against the San Antonio Spurs' Kawhi Leonard than against any other defender in the NBA.[26] It's not merely that Leonard is an excellent defender overall; it's that his specific *blend* of defensive skills presents an especially favorable matchup against James's offensive skills.

NBA players' day-to-day habits are also mined for anomalies, and this "behavioral" iteration of Moneyball is something entirely new. Jeremy Lin, a point guard with the Brooklyn Nets, thinks that his team's partnership with Infor has already started to pay dividends, by helping him to take care of his body just like a Formula 1 team takes care of its cars. In particular, he credits advanced analytics with improving his sleep and helping him recover more quickly from a nagging hamstring injury.[27]

Teams in other pro sports leagues have also embraced AI, for the same reason they've embraced ads on the jerseys: there's big money on the line. For example, Leicester City Football Club, in the English Premier League, made very clever use of player-tracking data during its title-winning season of 2015–16. The team had access to data from a system called Prozone3, which combines cameras and wearable sensors. Like all Premier League teams, it used that data to adapt in-game tactics to each opponent. But Leicester City also mined that data for something else: anomalies in a player's movements and workload that suggested an elevated risk of injury. These efforts led the team to the lowest injury rate and the most consistent starting 11 in the Premier League.

Postscript

We will leave you with one final detail about the Trial of the Pyx. It turns out that, even though English coins aren't made of silver anymore, the Trial still happens today. Every year on the second Tuesday in February, a jury of goldsmiths convenes in London to weigh a sample of coins and test their fineness. Luckily, they've learned from past mistakes: the bounds for declaring an anomaly have been calculated on a statistically sound basis since the middle of the nineteenth century.

There's one other difference, too: for the last 75 years or so, the jury has also tested the coins' width and diameter. Those particular measurements weren't important in Newton's day. Funnily enough, they're not that important today, either—for they were added to answer the demands of a fleeting time in the long history of England, when Londoners used coins to make phone calls from red boxes on the street corner.

6

THE LADY WITH THE LAMP

What the Crimean War can teach us about the prospects for an AI revolution in health care—and about the culture and institutions that help innovation take root.

IF YOU READ the news about health care these days, you'll encounter two very different narratives.

First, the bad news: health-care systems across the rich world are groaning under the burden of sick, aging populations. Obesity and heart disease are up, and costs are spiraling out of control. In 2016, two-thirds of all British hospital trusts ran a deficit, and the French health service outspent its budget by 3.4 billion euros. Americans, meanwhile, spend far more of their GDP on health care than anyone else but aren't any healthier because of it. Doctors pass their days fighting insurance companies, sweating lawsuits, and typing data into an electronic health-records system; compared to nondoctors, they are 40% more likely to abuse alcohol or drugs, and twice as likely to commit suicide.[1]

Perhaps as an antidote to all these depressing stories, we're also told that artificial intelligence is set to revolutionize health care. AI evangelists describe a futuristic world where your surgeon is assisted by a laser-guided robot, just like the Google car; where your vital signs are algorithmically monitored for anomalies, just like your credit card; and where your treatments are personalized, just like your Netflix account.

It's a world where your Fitbit can tell you whether you're going into labor, where you can snap a picture of a skin lesion and get an instant diagnosis from your phone, and where your smart watch knows just the right nudge to get you to eat more veggies or take the stairs.

In this world, doctors no longer spend a third of their time doing manual data entry. Instead, they tell everything to a sort-of Amazon Echo on steroids, which immediately updates your medical record—which is then analyzed using sophisticated prediction rules, trained on enormous databases, that help doctors look for hidden signs of trouble. It is a world of perfect interplay between human and machine intelligence, as cheap wearable sensors, coupled with AI-based diagnostic and monitoring technology accessible via smartphone, provide a significant upgrade in care for underserved communities—first in the rich world and then in the developing world. Childbirth becomes safer; diseases are caught earlier; vast oceans of human potential reach full tide.

We hope you agree that this world sounds pretty great—assuming we can address your concerns about data privacy, which we'll try to do by the end of this chapter. So our question is: Why aren't we living in this world already? All of the AI technologies we just listed already exist in some stage of research or development, and it's patently obvious what's needed in order to prompt their widespread adoption: better data, deeper collaboration between health-care providers and data scientists, and smarter laws that can foster innovation while safeguarding patients and their privacy. But as you'll learn in this chapter, just because something good *can* be done with data doesn't mean it *will* be done.

So far, we have highlighted examples of tremendous technological progress in AI. We'll now shift our focus to the interplay between technology and *culture*: the values, incentives, and habits that govern how people behave. To effect the kind of health-care revolution we all seek, we certainly need resources, data, and people. Above all, we need a cultural commitment—by doctors and nurses, hospitals, corporations, legislatures, and patients—to bring those resources, data, and people together. Google, Facebook, Amazon, PayPal, Baidu, Alibaba, Formula 1, the Mayor's Office of Data Analytics in New York, Makoto Koike's cucumber farm in Japan . . . they've all made this same commitment in their respective areas, with impressive results. Which makes it all the more tragic that such a commitment has been lacking in health care, where AI could help more people than just about anywhere else. We are likely still years away from seeing our most advanced AI technologies help real patients in substantial numbers, and the reasons have nothing to do with science or computing power and everything to do with culture, incentives, and bureaucracy. This isn't just a U.S. problem, either. Health-care systems in America, Europe, and Asia differ in important ways, but they share some important commonalities in terms of how AI could be helping, and why it isn't already. Cancer and kidney disease have no nationality, but there's a word for bureaucracy in every language.

At a moment like this one, it helps to seek a historical example of someone who faced a similar problem and overcame it—someone who possessed the knowledge, the stature, and the resolve to stand up to the powerful people running health-care systems and say on behalf of us all: Stop it, please. Why are you doing it this way? Can't you see how things could be so much better?

Luckily, we know just the person: Florence Nightingale.

Depending on your age, you may or may not know Nightingale as the most famous nurse of all time—the "lady with the lamp," who became a living symbol of compassion while tending the wounded British soldiers of the Crimean War. It turns out that when she wasn't caring for soldiers, Nightingale was also a skilled data scientist who successfully convinced hospitals that they could improve health care using *statistics*. In fact, no other data scientist in history can claim to have saved so many lives as Florence Nightingale. In 1859, in honor of these achievements, she became the first woman ever elected to the U.K.'s Royal Statistical Society.

Nightingale's path to unlocking the power of health-care data offers three distinct lessons for today. First, it illustrates the kind of institutional commitment necessary for a data-science revolution to take hold in a given field. Indeed, if you have any professional interest in how AI might change your field, you'll find no better lesson.

Second, it shows what you're up against as a patient who wants the best care. In her quest to bring better data analysis to health care in the 1850s, Nightingale fought entrenched interests that defended the status quo against changes that would help patients. The fight to do the same thing today is playing out in a shockingly similar way, and if the first time was tragedy, this second time looks like farce.

Finally, Nightingale's story is inspiring. Today's health-care system sure could use people with the tenacity, brains, and moral courage that Florence Nightingale showed 160 years ago. Maybe one of those people will be you.

The Angel of the Crimea

Florence Nightingale was born in 1820 into a life of comfort and privilege. Every year, her family would rent a hotel suite in town for the London season, before retiring to one of their two country estates. When they took European holidays, they traveled in a grand carriage with room for 12, and they enjoyed lavish entertainments: operas every week, "balls to infinity," banquets thrown by the Grand Duke of Tuscany.[2]

But Florence experienced this life as a gilded cage. For she had two real loves, neither of which could be found in the idle pleasures of the drawing room.

Her first love was mathematics. Even as a child Florence had thrown herself into her math book, solving problems from a vanished age: "If there are six hundred millions of Heathens in the world, how many Missionaries are needed to supply one to every twenty thousand?"[3] She played mathy word games—"I took 'breath' and I made forty words," she wrote at age seven.[4] As a teenager, she learned geometry from Euclid and logarithms from her cousin Henry, and she begged her parents to let her make an extended visit to her Uncle Octavius, who had a fantastic math library.[5]

Even more than math, Florence loved nursing. As a child she treated injured dogs, wrote an epitaph for a garden wren, and bemoaned the condition of a cow with a bad cough; as a teenager she visited the sick and the poor of the local village almost every day. When Florence went missing in the evenings, her mother knew to go knocking on doors in the village, where she could be found "sitting by the bedside of someone who was ill, and saying that she could not sit down to a grand 7 o'clock dinner."[6] If she *did* sit down to dinner, she was prone to asking awkward questions of any guest, no matter how distinguished, who appeared to turn a blind eye toward the suffering of others.

She soon set her sights on a career as a professional nurse, writing in her diary: "My mind is absorbed with the idea of the sufferings of man. . . . All that poets sing of the glories of this world seem to me untrue. All the people I see are eaten up with care or poverty or disease." She would awake as early as 3:00 A.M. to read anything she could find on social welfare: statistics from the census, minutes from Parliament, a "Report on the Sanitary Conditions of the Laboring Classes of Great Britain."

Her parents, alas, found her career ambitions frustrating and bizarre, completely unbecoming a lady of her station, and they refused her wish to enter a training program for nurses. Florence responded by dismissing their ideal of womanhood as one "fed on sugar plums," as the life of "the lark singing in the bright sunshine," never to "descend like the rest of us to the busy scratching rabbit warren, where the inhabitants are digging and burrowing and making a dust."[7] She felt guilty for being so unhappy, when she herself enjoyed such privileges while so many others led lives of penniless suffering. Yet as she reached her thirtieth birthday, and as her family stymied her wishes time and again, her thoughts had become depressive and suicidal.

Ultimately, though, Florence's power of will—what her sister Parthenope called "the most resolute and iron thing I ever knew"—triumphed. At age 31 she finally won her parents' permission to train in nursing at Kaiserswerth, a famous charity hospital in Germany. It was a turning point. Her stint at Kaiserswerth involved long hours with the sick and the hopeless. She dressed wounds, treated typhus, nursed amputees, and sat vigil at the bedsides of the dying. The experience left her feeling like a new woman. She was finally answering the call she'd been hearing her whole life—and as a friend wrote her, "You will find ordinary twaddling life more insupportable than ever after this taste of your own heart's choice."[8]

Indeed, when Nightingale finally returned to England, neither her family nor the polite conventions of her class would keep her from realizing her lifelong ambition. She began work at a small women's hospital on Harley Street in London, rapidly earning a reputation for skill and compassion, and in 1854 she was offered the job of her dreams: superintendent of nurses at King's College Hospital. But history had other plans for her—for that October, as the war in the Crimea between Britain and Russia heated up, Florence Nightingale would be called to serve her country.

"Foul Air and Preventable Mischiefs"

The spark for the Crimean War had been lit in 1853, when Russia invaded the Balkans, thereby threatening Britain's ally Turkey. In response, the British declared war on Russia in March of 1854, sending troops to the Crimean Peninsula to lay siege to Sebastopol, the main harbor for Russia's Black Sea Fleet. People back in London, full of nationalistic fervor, had assumed that the war would be over in a flash. These hopes of a quick victory were soon dashed, when it became clear that the British army, a generation removed from its last major war—against Napoleon, in 1815—was completely unprepared to face Russia.[9]

Nowhere was this more apparent than in the army's decaying medical system, where basic matters of sanitation and supply chains were thought beneath the dignity of the medical men in charge. The result of all this poor planning was a logistical and humanitarian catastrophe. A soldier wounded in the Crimea typically found himself packed onto a grimy ship and transported 300 miles to the Barrack Hospital at Scutari, opposite Constantinople on the Bosphorus. Upon arrival, he might wait as long as three days to be taken ashore, before being loaded on a stretcher, or maybe strapped to a mule, for a jarring climb up a steep hill to the filthy hospital. There he would encounter carnage, in the form of fellow soldiers sprawled on thin mats amid rats, blood, stench, and filth. Cholera and dysentery were rampant: the sewers were clogged, the toilets leaked excrement into the courtyard, and a water main was blocked by the decomposing carcass of a horse.[10] The hospital was badly short of medical supplies, clean clothes, healthy food, and chloroform—many amputations were performed without it.[11] Doctors were scarce, too, and all those who *were* around raced through the halls from one emergency to the next, dodging men and corpses.

By the autumn of 1854, heavy scrutiny had fallen upon the conditions at Scutari. A September 30 article in *The Times* channeled the public's growing outrage:

> Not only are they left to expire in agony, unheeded and shaken off, though catching desperately at the surgeon whenever he makes his rounds through the fetid ship, but now, when they are placed in the [hospital], where we were led to believe that everything was ready which could ease their pain or facilitate their recovery, it is found that the commonest appliances of a workhouse sick ward are wanting.[12]

Sidney Herbert, secretary of war and a close family friend of the Nightingales, came under enormous pressure. He had observed Florence's rapid rise in the field of nursing, and he approached her with a proposal. Would she consider leading a government-sponsored group of nurses to Scutari, to assist the doctors and tend to the suffering men?

Florence readily agreed, and she steeled herself for the worst. But nothing could have prepared her for the conditions she found upon her arrival: four miles of corridors filled with grotesquely injured men sleeping 18 inches apart, their lives immiserated by "foul air and preventable mischiefs." The hospital's supply chain, moreover, had broken down completely. Nightingale could find no linen to make bandages, nor fresh shirts to replace those soaked with blood. There was plenty of "gangrene, lice, bugs, and fleas," yet "no mops, no plates, no wooden trays, no slippers . . . no knives and forks, no scissors (for cutting the men's hair, which is literally alive), no basins, no towelling, no chloride of lime." She soon learned that requests for supplies needed to pass through eight different government departments back in London—and when these requests were finally processed, the wrong supplies were often sent, or else the right supplies were sent to the wrong place. At Scutari itself, Nightingale encountered only dawdling and obstruction from the purveyor-general. Matters were so bad that she asked *The Times* to entrust her with the donations it had collected for a soldiers' fund, so that she could bypass the purveyor and go shopping for necessities in the Grand Bazaar of Constantinople.[13] Thereafter, she effectively became the hospital's shadow purveyor, as the chief conduit for the enormous variety of gifts that ordinary citizens sent to Scutari—food, cash, linens, slippers, a drying cupboard . . . even raspberry preserves and ginger biscuits, from a Mrs. Gollop of Buckinghamshire.[14]

Although Nightingale was a gifted nurse, her talents shone even brighter as an administrator. She began modestly, by enforcing new standards of cleanliness, but she soon found herself charged with reorganizing virtually every nonmedical function at the hospital. Florence described her role as "cook, house-keeper, scavenger . . . washerwoman, general dealer, store-keeper."[15] The effort tired her to the bone. She worked 20-hour days and took meals on her feet. She was exhausted by "the quantity of writing, the quantity of talking . . . the dealing with the selfish, the mean." She felt "like Prometheus," bound to "the rock of ignorance [and] incompetency."[16]

Yet all the while she was making a difference. Only two months after her arrival, the hospital chaplain noticed a surprising "air of comfort and enjoyment." There were stoves on every ward and tin baths in every corner. Every man had a bed, a clean mattress, and a change of shirt twice a week.[17] And mortality was dropping: having peaked at a shocking 52% of admissions in the winter of 1855, it fell to 20% by March, and thereafter continued downward through the following winter, by which point it was no higher than the rate among civilians in a major city.[18]

Florence could hardly take all of the credit for this herself, and she never tried to.[19] Still, for more than a year, the medical operation at Scutari had been a ship barely surviving the gale—and, in the words of an army colonel who'd seen things firsthand, "Miss Nightingale [was] its only anchor." Her colleagues recalled her energy, her example, her way of cutting through red tape with a machete. They recalled the darkest days of winter, when wounded troops arrived by the hundreds and "the officials lost their heads—crying out to Flo" for this and that.[20] And they recalled the chaos that reigned during her short absences—like the one day in 1854 when she took a brief rest from her duties as unofficial purveyor, when the men of C corridor all ended up drunk, having guzzled their wine straight from the bottle, since no one had given them cups.[21]

Nightingale's Data-Science Legacy

Back in Britain, a journalist at *The Times* conveyed the image of Florence Nightingale that would endure forever: "When all the medical officers have retired for the night and silence and darkness have settled down up those miles of prostrate sick, she may be observed alone, with a little lamp in her hand, making her solitary rounds."[22] With time, her legend only grew. Poems and sentimental songs were written about her. Soldiers' private diaries recorded daydreams of leaping to her aid in the face of danger. Ships, racehorses, and babies of every social class were named in her honor.[23]

But to Nightingale, this reputation was nothing but a "false popularity, based on ignorance."[24] She believed that her work back in England, long after the war was over, ultimately made a much bigger difference—and modern historians largely agree with her. Three significant legacies stand out from that period, all made possible by the experience and fame that she gained during the Crimean War.

The Lady with the Lamp

Nightingale's first legacy was as a living symbol of nursing reform. Before her, the symbol of Victorian nursing had been Mrs. Sarah Gamp, the savage caricature of a domestic caregiver from Dickens's *Martin Chuzzlewit*. Untrained, uncouth, and perpetually drunk, Mrs. Gamp gave off "a peculiar fragrance . . . as if a passing fairy had hiccoughed, and had previously been to a wine vault." Her usual look, per Dickens, was "a leer of mingled sweetness and slyness . . . partly spiritual, partly spirituous, and wholly professional."

Mrs. Gamp was a stereotype, but her image became so iconic that it must have resonated with Dickens's contemporaries, for whom Mrs. Gamp became a symbol of the disgraceful state of nursing. As a prominent physician named Edward Henry Sieveking wrote in 1852: "Let the terms of nurse and gin-drinker no longer be convertible; let us banish the Mrs. Gamps to the utmost of our power; and substitute them for clean, intelligent, well-spoken . . . attendants upon the sick."[25]

In the wake of Nightingale's achievements, the public image of a nurse was transformed into more or less the modern notion we hold today. This could have happened only as a result of a decades-long period of reform in the training and certification nurses, and Nightingale was hardly the first advocate of these reforms. She drew inspiration from many earlier pioneers—especially the nurses who ran Kaiserswerth, where she'd trained in the early 1850s. Nonetheless, to the British public Nightingale became *the* symbol of the modern Victorian nurse. She did more than any other single person to make nursing a respectable path for middle-class women, thereby helping to create a virtuous cycle in which better nurses made for a better profession, attracting yet better nurses.

The Passionate Statistician

Nightingale's second legacy was her personal analysis of medical statistics from the Crimean War. Florence came home to England full of righteous indignation at the scandal of Scutari. In her diary, she wrote: "I stand at the altar of the murdered men, and while I live, I fight their cause."[26] It *was* a fight, against those in the army and the medical establishment who stood implacably in the way of change—like army doctor John Hall, for example, who dismissed Nightingale as a "petticoat imperieuse."[27] And she brought all her weapons to bear in that fight: her intellect, her network of friends, her acid pen . . . and above all, math and statistics, which she viewed as the mightiest arrows in her quiver.

Nightingale's first biographer, E. T. Cook, nicknamed her "the passionate statistician"—which for obvious reasons didn't capture the public imagination like "the lady with the lamp" but provided a far better description of how she changed the world for the better. Nightingale was especially adept at using graphical representations of data—"data viz," in modern parlance—to draw the nation's attention to the shameful conditions that had prevailed in military hospitals. As one of her colleagues put it, Nightingale's pictures of data could "affect thro' the eyes what we may fail to convey to the brains of the public through their word-proof ears." She even invented a new kind of statistical figure: the polar-area or "coxcomb" diagram, which showed changes in mortality over time using a series of colored wedges. Her coxcomb diagram from

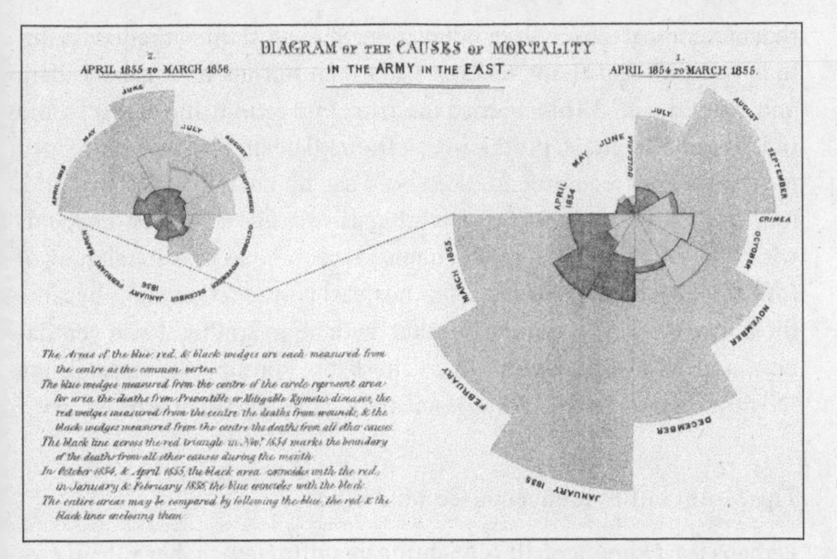

Figure 6.1. Nightingale's coxcomb diagram from 1858. In the circle on the right, the 12 wedges around the outside represent month-by-month deaths in the Crimean War from "Preventable or Mitigable Zymotic diseases" from April 1854 to March 1855. The dotted line then leads you to the circle on the left, which shows the data for the following year: April 1855 through March 1856. In each circle, the inner two sets of wedges represent deaths due to battle wounds (in black) and all other causes (light gray).

the Crimean War, depicting the rise and fall of mortality due to disease, is shown in Figure 6.1.

Her analyses revealed that in the first seven months of the Crimean campaign, British soldiers suffered a 60% mortality rate from disease alone. This was higher than Londoners experienced during the Great Plague of 1665, higher even than the probability that a civilian who contracted cholera in 1850 would die.[28] Yes, it was literally safer to have cholera at home than to take your chances in the Crimea—and that was *before* you faced a single enemy bullet. Nightingale referred to this as "the finest experiment modern history has seen . . . as to what given number may be put to death at will by the sole agency of bad food and bad air"—an experiment that condemned 16,000 men to death.[29]

She also analyzed peacetime statistics and discovered that, owing to poor sanitary conditions, the army's rate of mortality at home was twice that of a comparable civilian population. She called this situation "criminal," no different than "to take 1,100 men out upon Salisbury Plain and shoot them."[30] This shamed the army into retrofitting barracks and redesigning hospitals, producing an immediate drop in disease-related mortality.[31] Her recommendations soon caught on in the civilian world, too. Due in no small part to Nightingale's tireless advocacy, hospitals with long corridors and stuffy rooms came to be seen as incubators of infection. Her preferred model of hospital construction soon became the norm: the pavilion-style hospital, with abundant light and ventilation, and with separate wings to check the spread of disease. These "Nightingale wards" remained popular long into the twentieth century.[32]

The Mother of Evidence-Based Medicine

Perhaps least known of all is Nightingale's third legacy: her role in creating a new standard of professionalism in the collection and analysis of medical data.

It is often said of generals that they are always fighting the last war. But an army doctor looking for lessons in the enormous variety of medical experience from the Crimean War wouldn't have been able to do even that. No statistics were collected, few clinical histories were preserved, and almost no postmortem examinations were done. In many cases, sick men were loaded on one side of the boat in the Crimea, only to be thrown overboard dead from the other side when they reached Scutari. Nightingale despaired at the fate of the men, but she also found it "discouraging and disappointing in the extreme" that such a "scientific treasury" had been lost to mismanagement.[33]

Upon returning to England after the war, Nightingale discovered that these failings were mirrored in the civilian world. The country had no system for the collection of even the most basic medical statistics, like recoveries, lengths of stay, or mortality from different diseases. Even if there had been such a system, there would have been no way to compare results across different hospitals, which all used idiosyncratic classification systems for disease.[34]

Nightingale saw this lack of attention to data as a public health emergency. She saw how the new discipline of statistics was transforming other fields, like astronomy and earth science. She also noticed how Continental statisticians—most notably the eminent Belgian Adolphe Quetelet, one of her idols—were using these new tools to address complex social-science questions about crime and demographic change. Nightingale saw *incredible* potential in applying those same statistical techniques to health care; it "would enable us to save life and suffering, and to improve the treatment and management of the sick."[35] But that required much better data from the health-care system. To that end, she drew up a standard set of medical forms, obtained the endorsement of many of the world's leading statisticians, and urged the big hospitals in London to begin using them. She also lobbied the government to begin collecting data on illness and housing quality as part of the census, arguing that "the connection between the health and the dwellings of the population is one of the most important that exists."[36] From top to bottom, Nightingale's work clearly foreshadowed the coming 160 years of evidence-based health care. Her ideas formed a clear model for the international system of disease classification used today, which serves as the bedrock for all of modern epidemiology and medical data science.[37]

Preventable Mischiefs in the Age of AI

Nightingale's three legacies all have clear parallels today. They also raise some sharp questions. She spoke of the "foul air and preventable mischiefs" that killed the soldiers of the Crimea, and while the air in modern hospitals may be less foul, there are still mischiefs aplenty.

One big question is how to staff and train a modern health-care team. After Nightingale, no hospital could function without nurses. When will the same be true of data scientists and experts in artificial intelligence, who now play almost no day-to-day role in health care?

A second question is how to design a hospital for this new age. Nightingale helped to establish new sanitary standards, and hospitals were reengineered from the ground up in response. When will hospitals undergo another epoch of redesign, to accommodate what is now possible using artificial intelligence? When will data hygiene be taken just as seriously as patient hygiene?

Finally, the most important question of all is how medical statistics should be collected, shared, analyzed, and used. We've improved a lot in that department over the last 160 years, thanks in no small part to Nightingale's efforts. As you'll soon learn, though, we've improved only in certain ways—and we could be doing so much more. In light of what's happening *outside* of health care, this is starting to look like a moral embarrassment. We live in an age when Formula 1 cars are monitored in real time by algorithms and teams of engineers, when your movie-watching preferences are the concern of multibillion-dollar AI operations, and when your propensity to click on an ad for dog food is analyzed on supercomputers, using millions of variables and billions of data points. Yet for the most part, we still rely on numbers that Florence Nightingale could have crunched with pen and paper to quantify the risk that your kidneys will fail. And in some ways we haven't improved at all: a 2017 paper in the *Journal of the Royal Statistical Society* referred to Nightingale's 1860 protocol for hospital data collection as "conceptually more complete" than many systems today.[38] Which should leave us all wondering: When will medical data science move into the twenty-first century?

Why We Need AI in Hospitals

We want to be clear up front that this *isn't* the fault of individual doctors and nurses. Rather it's the fault of the whole health-care system, which for too long has been the Mrs. Gamp of data science: backward in its statistical norms, drunk on bureaucracy, and ignorant of what modern AI has made commonplace.

To illustrate this point, we'd like to tell you the story of a man from somewhere on the East Coast of the United States—let's call him Joe—who died at age 62 with chronic kidney disease. Joe's story explains a lot about how the contemporary approach to medical data science is failing patients, and why a combination of better data curation and AI could prevent so much suffering.

By his midforties, Joe was already suffering from type 2 diabetes and congestive heart failure. Maybe his job was stressful, or his diet and exercise habits poor. Whatever the mix of causes, they finally caught up with him. A few weeks shy of his forty-seventh birthday, Joe felt a sudden numbness in his right arm. He stumbled and fell heavily to the ground. He was taken to the emergency room, and he was immediately diagnosed with an ischemic stroke, meaning that a clot had blocked the flow of blood to his brain.

Fortunately, Joe survived the stroke. Although his high blood pressure and diabetes marked him as having a higher risk for kidney disease at some point in the future, for now his kidneys tested fine. The standard measure of kidney function is the GFR, or glomerular filtration rate. Joe's GFR was estimated at 99, well above the danger zone: a GFR of 60 or below indicates a mild to moderate loss of kidney function, while 30 or below means severe loss.[39]

Over the next year, Joe made nine more trips to the emergency room for maladies of one kind or another, none of them explicitly related to his kidneys. On two of those occasions, he was admitted to the hospital, and his kidney function was measured: his GFR was first 96, and then 95 about a month later. This decline was a little bit steeper than the expected rate of 1–2% per year in a healthy person. Still, each individual reading was above the clinical threshold of 60 that usually worries doctors.

About a year after his stroke, Joe started making regular trips to an outpatient clinic: eight visits over 14 months. On every visit, the doctor ordered a routine series of tests, and the clinic staff duly entered data on Joe's kidney function into an electronic database—the same one used by doctors at the hospital. His GFR numbers yo-yoed a bit between 60 and 75: still above the threshold of 60, but down quite a bit from the prior year's reading of 99, and on an unmistakable downward trend.

At age 49, Joe was readmitted to the hospital, and his GFR was measured at 54. Over the next several months, he made 10 more visits to the ER, as well as a dozen more visits to the clinic. Joe was now very sick. A month before his fiftieth birthday, his GFR was measured at 40, well into the danger zone. Yet he received no treatment that might have prevented his slide toward kidney failure. We can only speculate why, but one reason might be that test results sometimes take a while to come back from the lab—by which point the patient might be already home, and no longer under the direct care of the doctor who ordered the test in the first place.

Over the next three years, Joe had 20 further encounters with a doctor. On many of these occasions, his kidney function was measured, and it was dropping at a scary rate: below 30 by age 51, and below 20 by age 52, whereupon Joe was finally referred to a kidney specialist, more than a year after his GFR readings fell below the level that typically triggers such a referral.

But kidney failure was now inevitable. Three months after his appointment with the specialist, Joe's kidneys finally gave out. He was rushed to the emergency room, his twenty-fifth visit since his first stroke. His GFR was measured at 12; his kidney function had declined 34% per year, each of the last five years, from his initial reading of 99 after the stroke. The doctors in the ER put him on emergency dialysis, one of the most traumatic and expensive procedures on the books.

For the next decade, Joe became what the insurance industry refers to as a "super-utilizer," which is management-speak for an appallingly sick human being: one of the 5% of patients who account for more than 50% of all health-care spending in the United States. In Joe's case, this meant severe diabetes, stage 5 kidney disease, angina, vascular disease, and inflammatory connective-tissue disease, along with a series of heart attacks. Joe's kidneys were tested 124 times over this period, which included 26 more visits to the ER and nine to a kidney specialist. His GFR bounced around, but it never again rose above 20.

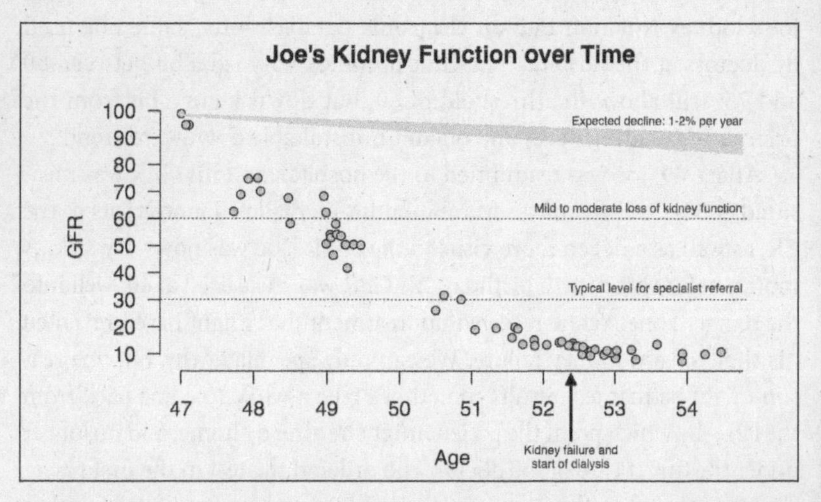

Figure 6.2.

He died a week shy of his sixty-third birthday, roughly 10 years after beginning dialysis.

What did Joe die of? In one sense, the answer is clear: his kidneys failed. But in order for that to happen, something else had to fail first, in a manner so complete that it beggars belief. For if you take all of Joe's GFR readings in the eight years after his stroke and you plot them over time, the trend looks pretty obvious (see Figure 6.2).[40]

In those three years of steep decline between ages 47 and 50, not one of Joe's health-care providers had looked at a simple plot of his GFR readings over time. The issue was, quite literally, one of failing to connect the dots. Doing so would have yielded a simple and obvious prediction: this guy's kidney function is declining so rapidly that it will probably keep declining, and if it does, the result will be painful and expensive.

Joe certainly died for want of a kidney. But more fundamentally, he died for want of a scatter plot.

Threshold Thinking

How could this have happened? We posed this question to Dr. Katherine Heller, a professor of statistics and machine learning at Duke University, who analyzed this data and brought Joe's case to our attention in the first place. "In retrospect," said Heller, "that steep decline between ages 47 and 50 represents such an obvious missed opportunity. All you have to do is draw a straight line through the cloud of data points, and you can see where things are going."

So why did no one, whether human or machine, draw that straight line? *This* is the essential question in modern health care. To understand the answer, we must revisit two earlier questions that Florence Nightingale asked 160 years ago, when she pondered how the new mathematical tools of the 1850s might be used in hospitals:

1. How does the health-care system use data today?
2. In light of new data-analysis technologies, what *could* it be doing instead?

Today, the main way the health-care system uses data is to create checklists. These checklists encode the "standards of care" recommended by national bodies, like the American Medical Association or the U.K.'s General Medical Council. These standards of care, in turn, are ultimately driven by data from published research findings: about what warning signs to look for, about what treatments actually work, and about what diagnostic protocols help the most people. For example, you may recall that the American Cancer Society created some controversy in 2015 when it updated its recommended checklist regarding mammograms. According to the new checklist, women at average risk of breast cancer should start getting annual mammograms at 45, rather than 40. That change was made only after a team of 19 experts conducted a massive synthesis of all available data and concluded that the new recommendations were likely to prevent 11 false positives out of every 500 women screened, with no discernible impact on the number of deaths from breast cancer.[41]

Medical checklists are great, and the manner in which they're created and updated represents a triumph of data over anecdote—something that Florence Nightingale, were she still around, could take immense pride in. Checklists save lives by helping doctors catch subtle clues when making complex decisions. Inspired by his experiences as a surgeon, the medical writer Atul Gawande has even written *The Checklist Manifesto*, about how checklists can help in making complex decisions *everywhere*, not just in medicine. He makes a good case.

But checklists can fail—especially when they rely on what Katherine Heller calls "threshold thinking." To see this, let's return to the trend that's so obvious from the scatter plot of Joe's kidney readings that we showed you earlier. Heller surmises that each doctor along that sad trail of dots was thinking about Joe's case in terms of a binary threshold on a checklist. Is the patient's GFR above 30? Check. Are his levels of blood potassium below 5.5 millimoles per liter? Check. Does he have normal levels of albumin in his urine? Are his other kidney-related indicators within their expected ranges? Have I followed the protocol? Check, check, check.

All those "checks" tell you something about Joe's kidney function on that one isolated visit, and they're really important for delivering good health care. But those checks don't tell you anything about the long-term trend. So even though Joe had been hurtling toward that terrible GFR threshold of 30 for years, he hadn't yet crossed it—and no one raised an alarm until it was too late. In retrospect, this shouldn't be surprising. In AI terms, checklists are just prediction rules: procedures that take patient data as an input and produce a clinical decision as an output. As prediction rules, however, they are designed to help doctors understand and respond to what's happening *right now,* not what's likely to happen in the future. In fact, that's an inherent design feature of checklists: they focus the doctor's mind on the details of the present. But in a world where the biggest and most expensive medical problems are chronic diseases that unfold over a time span of years, that feature is starting to look like a bug.

You might ask: Why not just fix the bug with a longer checklist, by adding an item that encourages doctors to look at the long-term trend? We had the same question. So we asked Heller whether it would have even been possible for someone at Joe's bedside to have called up his GFR readings on a screen and plotted them over time, to look for a trend. "Maybe you could query the database that way, if you knew how," she said after a few moments' thought. "But it would certainly not be a natural and obvious way for a doctor to use the system." To see the trend, she continued, "you would really need to go back through the record manually, one reading at a time." Ironically, this probably would have been easier back in the days of paper charts.

Moreover, Heller pointed out, it's not just one set of readings to look at but hundreds or even thousands: blood tests, urine tests, EKG, heart rate, blood pressure, clinical symptoms, social factors—and soon, information on a patient's gene expression and epigenetic profile. There's just *so much data.* It's hard for a human being to comprehend it all even as a single snapshot, much less as a story that unfolds over time.

Finally, there's the issue of how such a hypothetical "look for trends" item on a checklist would fit into a doctor's usual workflow. When you show up to the emergency room, your doctor's main concern is: How bad is your case right now? Should you be treated and sent home, or are you sick enough to be admitted to the hospital? Doctors face high stakes and enormous pressure in making those decisions—and even outside of the ER and in a normal clinic, they have to make them *fast,* because there are dozens of other people in the waiting room who need their help, too. How reasonable is it to expect those doctors to stop what they're doing, fire up a statistical software package, and mine through a vast collection of electronic health data—all to find the one or two historical trends that might be relevant for preventing something months or years in the future?

Dr. Mark Sendak, of Duke University's Institute for Health Innovation, explains that while doctors might do this kind of thing on TV shows like *House*, they don't do it in normal hospitals. "Physicians always say they want the data," Sendak says. But, he continues:

> The problem is that there's no workflow for them to access or use the data. The way that the records are structured, it takes time and skill. You have to write a query, you have to download the data into a spreadsheet, and then you actually have to do things with it. But physicians are already under so much stress. They have 15-minute clinic visits. When, exactly, are they going to be playing with the data for their clinic and figuring out what they need to do for their patients?

That brings us to an even deeper issue: the entire system of medical data science was designed *only* to address questions at the level of a population. For example, how many lives would we save if we used threshold A rather than threshold B for detecting kidney disease? There must be hundreds of studies that bear on any such question. But medical data science is nearly silent in response to basic statistical questions at the level of an individual patient. How are Joe's GFR readings changing over the long term? Where are they likely to go from here? What does that predict for Joe's health next month, or next year? These questions would have been straightforward for either a person or an algorithm to answer, using Joe's historical medical record—yet all those data points were never given a chance to speak. There was no routine in place to sift Joe's health record for signs of an underlying chronic condition: no team of data scientists, no algorithm, no doctor with interdisciplinary training in statistics.

With some exceptions here and there, the same is true at most hospitals and clinics. In speaking with friends and colleagues about this topic, we've noticed that many people have the impression that there must be some medical "robot car" behind the scenes of a modern hospital—some fancy suite of algorithms that analyze patient-level records and help doctors make personalized recommendations and decisions. Perhaps they get this impression from seeing their own doctors doing so much data entry or from seeing AI transform so many other industries. Whatever the reason, they're usually shocked when we tell them the truth: that when it comes to patient-level data analysis at most hospitals today, not only is there no robot car, there is literally *nobody at the steering wheel.*

When we spoke with Heller, her frustration on this point was obvious. "It turns out that it's not enough to just collect all that data," she put it wryly. "You actually have to do something with it." Here she unknowingly channels Nightingale, who wrote of St. Thomas's Hospital in London, in 1859, that it "appears to keep its statistics more for the sake of checking obstreperous patients, which is an object certainly, but not a scientific one."[42]

The story of Joe, it turns out, is much more than a story of a man with kidney disease. It is a story of the vast canyon between what data *could* do for us and what our health-care system *lets* it do.

AI to the Rescue?

If it seems to you that health-care professionals are drowning in data and could really use a life preserver—that a combination of human and machine intelligence could radically improve health care—then you're not alone in that thought. Companies and researchers are hard at work on a new generation of AI-based technologies that stand waiting in the wings, ready to help doctors and nurses do their jobs more effectively.

Dr. Katherine Heller's team at Duke, for example, has teamed up with physicians to develop an AI system that can flag signs of impending chronic kidney disease.[43] At the core of their system is a prediction rule, just like the kind we met in chapter 2: it examines a patient's historical GFR readings, combines them with the data from other lab tests and vital signs, and makes a prediction for the future trajectory of that patient's kidney function. That prediction is shown in a mobile app that doctors can call up while treating a patient. With this kind of AI, doctors *really can* make the long-term trend part of their checklist, without wading into the data themselves.

Similar early-warning systems for other conditions have been invented by other research groups—for cardiac arrest, for depression, for fetal distress during labor, and for hospital-acquired infections, to name a few. Other equally impressive advances in AI technology could soon revolutionize every area of medicine, from radiology to cancer care to dermatology. We'll dig a bit more deeply into the cutting edge of the field, before returning to the question of what cultural changes must take place before we see widespread adoption of AI in health care.

Smart Medical Devices

The electrosurgical knife, which uses high-frequency radio waves to heat tissue to the point where it vaporizes, is a big improvement on the surgical scalpels of yore. The knife allows for drastically more precise cuts—and because it cauterizes the surrounding tissue almost immediately, it also minimizes blood loss. But even the fanciest knife can hardly be expected to help the surgeon know *where* to cut. When cancer surgeons remove a tumor, for example, they often find it impossible to tell by eye exactly where the tumor ends and healthy tissue begins.

Remarkably, a new AI-based smart knife, developed by Dr. Zoltan Takats and his team at Imperial College London, may soon help them. When tissue is vaporized by an electrosurgical knife, it creates smoke, which is normally sucked away by an extractor. Takats, however, had a clever realization about that smoke: it should contain metabolites from the vaporized tissue, which can be used to infer whether the tissue was cancerous. So he built an electrosurgical knife in which the smoke is instead routed to a mass spectrometer, which builds up a chemical profile. This profile is then fed into a prediction rule that classifies the smoke as having come from healthy cells or tumor cells. What's more, this four-step process—vaporize, extract, profile, classify—all happens in three seconds or less. As a result, the new knife actually *can* tell surgeons where to stop cutting. And in a trial involving real surgery patients, the knife's AI software identified the correct tissue type 91 times out of 91, as verified by postoperative histology.[44]

Some smart medical devices can even go beyond measurement into the realm of automatic treatment. Take the closed-loop artificial pancreas, an AI system that mimics the hormonal function of a real pancreas by automatically giving diabetics the correct amount of insulin in response to changes in their blood sugar. An artificial pancreas involves three steps: measurement, dosage, and delivery. The measurement step uses a continuous glucose monitor, or CGM, which provides a 24/7, real-time measurement of blood sugar. The next step is a dose-calculation algorithm: a clever prediction rule takes all the real-time blood-glucose data from the CGM and outputs the appropriate dose of insulin. The final step is a pump that provides insulin as needed.

Medical-device companies like Medtronic, Insulet, and Tandem are making fast progress in this area—and in a good omen, regulators are actually keeping up with them. In September of 2016, for example, after a surprisingly rapid three-month review period, the U.S. Food and Drug Administration approved Medtronic's latest model, the first artificial pancreas designed to level out both high and low blood-sugar levels.[45]

AI for Medical Imaging

Diagnostic imaging offers an even more immediate example of where AI can make a difference. Many common forms of medical image analysis, from looking at a chest X-ray to examining cancer cells under a microscope, involve a classic pattern-recognition problem: the inputs are the features extracted from the image, and the output is the diagnosis. As we learned in chapter 2, computers are absolutely brilliant at learning how to predict outputs from inputs, especially for images—and with more data and newer pattern-recognition algorithms, they're getting better all the time.

For some image-driven diagnoses, you soon may not even need to visit a doctor's office. Take, for example, the problem of diagnosing a skin lesion. The stakes here are high: melanoma causes more than 10,000 deaths per year in the United States alone. The five-year survival rate for melanoma is over 99% if it's detected early, but that figure drops to 14% if detected much later. For a number of reasons—time, money, a general aversion to doctors—people often resist going to the dermatologist until it's too late.

But a 2017 research article in *Nature,* by an interdisciplinary team of scientists led by Sebastian Thrun at Stanford University, described an AI system that may soon provide access to a vital bit of diagnostic skin care, for free, to anyone with a smartphone. The Stanford team knew a lot about algorithms for image recognition from their prior work on robot cars, and this gave them a simple idea. Instead of training these algorithms to distinguish a stop sign from a deer crossing, as they do for a robot car, what if we trained them to distinguish one type of skin cancer from another, based on an ordinary photograph?

Theirs was hardly the first effort at computer-aided dermatology, but the Stanford team made three crucial choices that separated their algorithm from a large pack of other, far less successful approaches. The first was scale. Previous efforts at doing this sort of thing had used small data sets, with fewer than a thousand images of skin lesions. The Stanford researchers compiled 19 databases containing 129,450 images, each of them classified according to a taxonomy of 2,032 different skin lesions. More data means a wider range of experience and thus better pattern recognition, like a veteran dermatologist who's been looking at skin lesions for decades and who's seen it all.

The second choice was their approach to computer vision, which involved the deep neural networks we met in chapter 2. These networks can extract subtle visual features, and they can combine those features into high-level visual concepts—like circles, edges, stripes, texture, or nuances of variegation—that can be used to distinguish 2,000 different types of skin lesion. They can do this, moreover, without ever being told by a programmer what to look for.

The final choice the Stanford team made was to use images from ordinary cameras, rather than highly standardized medical images that can be obtained only from a biopsy, or using equipment that only a dermatologist would have. These images exhibited wide disparities in lighting, color balance, zoom, and angle—irrelevant sources of variation that can easily fool a lesser algorithm into hallucinating differences that do not exist. But what these images lacked in quality, they more than made up in quantity; it would be much harder to collect 129,000+ standardized images from a dermatology clinic.

The result of all this work was an end-to-end AI system that, from an ordinary photo, can make two crucial inferences about a skin lesion. It can distinguish the two most common types of skin cancer from each other, and it can also distinguish a benign mole from the deadliest form of skin cancer: malignant melanoma. It can do so, moreover, with accuracy comparable to a panel of 21 board-certified dermatologists. On some measures, the Stanford algorithm even performs a little better.*

Similar image-analysis techniques will soon touch every area of medicine, as new specialties like computational radiology and computational pathology reach maturity. One research lab at ETH Zurich, for example, has developed an AI algorithm for grading the severity of inflammatory bowel disease from an abdominal MRI.[46] Another lab at Memorial Sloan Kettering Cancer Center has built a system for classifying renal-cell carcinoma from digital microscope slides.[47] And Moorfields Eye Hospital in London recently partnered with Google DeepMind to analyze over a million images from eye scans. The result was a neural network capable of automatically detecting signs of eye disease, like diabetic retinopathy and macular degeneration.[48]

Hardware companies have also responded to the exploding demand for AI-powered medical imaging. The chipmaker Nvidia, for example, is mostly known for its high-end computer graphics cards (GPUs) for gamers and filmmakers. But its hardware is also highly coveted by researchers in artificial intelligence who work with images and video. Recognizing this, Nvidia recently started building GPU-powered supercomputers that come bundled with software designed explicitly for medical image analysis. Massachusetts General Hospital was one of its first clients, and Nvidia now wants to train 100,000 new software developers to use the system for AI-based imaging.[49]

* The caveat is that both the doctors and the algorithm were making judgments from pictures alone, which is a bit artificial; you'd expect both to do better with more clinical information about the patient.

Remote Medicine

The phrase "remote medicine" conjures images of people living in far-off places with limited access to health care, like a spaceship or a North Sea oil rig.[50] For many people, however, medicine remains remote not merely for reasons of physical isolation. Think of the hundreds of millions of people living in the developing world, or the tens of millions of disadvantaged Americans who fall between the cracks of the private and public insurance systems. Think, even, of the ordinary middle-class person who has a job and a busy family life, and who just doesn't like going to the doctor.

AI-based remote medicine promises each of these groups a significant improvement in health care. Imagine a version of the Stanford team's skin cancer detection algorithm for a whole range of diagnostic problems. Think of hooking up a cheap stethoscope to your phone, so that a neural network can listen to your heartbeat. Or of staring into the camera to allow an algorithm in the cloud to scan your eyes for symptoms of eye disease. Now think of putting those algorithms together with something like Dr. Alexa: a digital assistant trained on vast troves of medical knowledge that's been programmed to ask questions about your symptoms and respond appropriately. (IBM's Watson team has already developed something very much like this for the purpose of training medical students.[51])

A new generation of wearable sensors could boost the effectiveness of AI-based remote medicine even further. If you think your Fitbit is cool, wait until the first person in your office gets a biometric e-tattoo: a small wearable patch, with the same thickness and elasticity as human skin, that can send health data wirelessly to your phone. These "epidermal electronics" are like a Formula 1 monitoring system for your health. They can measure your blood pressure, muscle strain, hydration level, respiratory rate, even the electrical activity of your heart or your brain—and they can immediately flag any anomalies. Such a system could be used by doctors to monitor someone just discharged from the hospital, or by ordinary people who want to track their health as they go about their day-to-day lives.

These technologies won't replace sophisticated laboratory diagnostics, and they certainly won't replace in-person care by a highly trained doctor. But for a nontrivial range of conditions, they could recommend simple treatments and funnel you to a doctor if and when you really needed one, all at very low cost. This kind of first-line-of-defense, AI-based diagnostic care, could vastly extend the reach of doctors and help them to treat problems long before they fester into something deadly and expensive—a perfect marriage of human and machine intelligence. The implications for medicine in the developing world could be especially dramatic, by lowering the cost of monitoring technology and making it far more mobile.

What Happens Next?

We hope you agree that all of this feels pretty exciting. Still, beyond the fact that some of these technologies are in early stages, there are many other cultural barriers that must be addressed before we expect to see them in widespread usage.

Incentives

To illustrate these barriers, let's return to the example of the AI-based early-warning system for kidney disease. Would hospitals even buy in? According to Dr. Mark Sendak, the question that every hospital will be asking itself is: "What does it mean for my bottom line if you can better predict kidney disease?" You don't have to be much of a cynic to observe, as Sendak does, that "huge health systems make money on advancing chronic disease."

The issue of incentives isn't specific to America. All countries, even those where the government pays for health care, face the daunting problem of making sure that everyone in the system is both motivated and empowered with a view to the long term. Sendak echoes this point: "Part of improving data science in health care is about aligning incentives, so that everyone cares about what happens to their patients when they're not in the hospital." That way, doctors will pressure their bosses to give them the tools they require to make decisions with the patient's long-term interests in mind. Right now, that's not happening;

if you're incentivized only to care about patients when they're in front of you, Sendak argues, "then you don't care how the data is stored, and you don't care about analyzing historical records to find patterns."

The legal system provides another set of incentives—or rather, disincentives. Imagine being in Dr. Katherine Heller's position, as you ponder the wisdom of commercializing, or even just giving away, an AI-based app that can forecast the progression of kidney disease. That app could probably help a lot of people, but it might also pose enormous legal peril for its designers. It's not clear whether the app makers, the data scientists, or the doctors using the app—or all three—would be liable for a bajillion-dollar judgment from the first inevitable missed case of kidney disease, regardless of how many lives the app saved, and regardless of whether its medical advice had all the appropriate caveats. That's because lawyers and policy-makers haven't gotten off their backsides to address a basic question: Who, ultimately, is responsible for an algorithm's medical advice? How can we answer this question in a way that simultaneously fosters innovation and protects patients?

Data Sharing

That bring us to another big question: Will data-science teams get access to the data they'll need to improve the existing AI systems and build new ones? If you work for a single hospital, you might have access to thousands of patient records. But wouldn't millions of records from lots of hospitals be much better? After all, a big reason that tech firms like Google and Facebook have such good AI is the sheer scale of their data sets. There are surely millions of clinical histories of kidney disease scattered across the medical databases of the world. In principle, these could be brought together, and teams of data scientists could be hired to analyze them using cutting-edge AI tools, in a way that still ensured patient privacy. Doing this across all of medicine would create hundreds of thousands of jobs, together with immense sources of social and financial value.

Yet there is little chance of that happening anytime soon. First, American health-care providers lack a common standard for their electronic records, making it effectively impossible to pool data and achieve the scale necessary for our best AI techniques to work their magic. Even in countries with centralized health systems, like the United Kingdom, the interoperability of databases—for example, between those used by general practitioners and hospitals—remains a huge problem.

Second, even if there were a common data standard, most hospitals are very hesitant to team up with data scientists, even under terms that guarantee patient privacy. In fact, we've found them to be downright paranoid, and other researchers we've spoken with have said the same thing. American hospitals tend to view their data—*your* data—as a closely guarded corporate secret. Nobody will really tell you why, but we've always suspected that it's for a pretty craven reason: hospitals don't want their Byzantine pricing models to be reverse engineered by their competitors, and so their default corporate position is to simply lock up the hard drives. Whatever the reason, all those electronic health records are used to generate very detailed bills, but almost never to help people require fewer expensive hospital services in the first place.

We find this mind-boggling, and we're not alone. Can you imagine if we allowed hospitals to treat organ donation the same way? What if hospitals could hoard your kidneys when you die, just like they can hoard the data about your kidneys? Shouldn't there be a form you could sign to overrule them, donating your data to save someone else's life? As Sendak puts it, "The moral imperative is that these are people who pay us for health care. We collect their data, we charge them, and we do nothing useful with their data. If we are the only people with this information, and they're paying us to take care of them, how are we not using it?"

That brings us to the issue of medical data itself, which is typically full of errors and missing entries—in other words, pretty much exactly what you'd expect if you asked a bunch of harried doctors to type in data manually, as fast as they can between appointments, after you've made it clear to them that most of the data will never be used to do anything useful. So when some lone research team comes along and gets permission to use some tiny fraction of that data to build an AI tool, they must first clean and curate the data. That requires skill, patience, and teamwork with clinicians—and right now, those collaborations are unscalably ad hoc. Imagine one team of researchers spending six months cleaning a data set with tens of millions of data points, all to ask one specific question—say, how to predict the progression of kidney disease—and to publish two academic papers. There is no clear way for others to benefit from that data set; nor is there any system to facilitate similar interactions at a large scale. Imagine if you had to write your own GPS mapping software every time you wanted to hail a ride on Uber or Lyft. You'd probably just take a taxi.

And the hospitals themselves, again with some exceptions, don't seem to be rushing to hire their own in-house data-science teams. The result is a sad misallocation of talent. The best data scientists of our generation could have been working on health care for years now. Many would love to, and to give away the wonders they create. Instead, they've been thinking up better ways to make you click on ads—because that's where the data is.

Privacy

The next question is about the privacy of your health information. This is a big issue that we can't treat in its entirety, but one important fact to highlight is that we're talking about health-care data that hospitals already collect, and that hospital staffers already access in order to send out bills. Building an AI system entails hiring someone to analyze this preexisting data on-site, or else giving an external data scientist remote access on a secure server, after having removed all identifying information.

That fact comforts many people, but it certainly doesn't address every *possible* concern about privacy or security. For example, you might worry that some malicious data scientist actually *could* identify people based on their health records, even from a supposedly de-identified data set. In fact, of all the problems we've listed here, this is the only one that can actually be addressed by new technology—specifically, by something called "differential privacy." Statisticians and machine-learning researchers think *a lot* about data privacy, and they've invented a variety of data-analysis tricks—mathematical techniques with names like "subsampling," "cryptographic hashing," and "noise injection"—that keep each individual person's records completely secure. With these new differential-privacy algorithms, a hospital data-science team could store health-care data in a way that lets them build accurate prediction rules and yet still prevents a rogue actor from learning some specific privacy-violating detail about any one patient. Suffice it to say that most hospitals are nowhere near implementing these kinds of algorithms, but the algorithms themselves certainly exist. You can even find them on any new phone that runs iOS or Android—where, for example, they're used to analyze which autocorrect suggestions you overrule in text messages, while simultaneously keeping the messages themselves encrypted and secure.

Then there's the issue of hacking. Hacking already plagues hospitals: if you recall the big ransomware attacks of 2017 (like WannaCry), you may also recall that hospitals were disproportionately hit. These hospitals probably weren't doing anything AI-related with their data, but that kind of activity would hardly have entailed a higher security risk than what was already present. Hospitals should obviously plug their existing information-security holes—probably, as many experts suggest, by moving to some kind of cloud-based infrastructure run by a firm who thinks about security full time. But this has nothing to do with whether the data already sitting on hospital servers should be used to improve health care.

Postscript

As you now appreciate, when it comes to widespread adoption of AI, the health-care system faces very few barriers of technology, but enormous barriers of culture, law, and incentives. Some of these barriers are specific to America, but many others affect all health-care systems of the rich world.

The upshot is that the next data-science revolution in health care won't take just one person like Florence Nightingale but thousands of them. It will take researchers like Katherine Heller and Zoltan Takats and Sebastian Thrun and Mark Sendak—people who keep working on cool projects, keep convincing their colleagues inside health care that this AI stuff really works, and keep generating good evidence. It will take doctors, nurses, software engineers, lawyers, database managers, privacy experts, venture capitalists, insurers, hospital administrators, policy-makers, and patients, all of whom must come together to make this thing work.

May Florence's power of will—that "most resolute and iron thing"—live on in them all.

7

THE YANKEE CLIPPER

Baseball, big data, and the importance of assumptions.

THERE'S A PECULIAR strain of AI evangelist who believes that smart machines will soon make people irrelevant to the process of discovery. Before long, the story goes, we won't need theories or assumptions to learn about the world. All we'll need is the right deep-learning algorithms, turned loose on the right data sets, and we'll end up with so much knowledge that it'll be pouring out of our nostrils.

People have been making these kinds of predictions for a while. In 2008, for example, the editor in chief of *Wired* wrote: "Science can advance even without coherent models, unified theories, or really any mechanistic explanation at all. . . . We can throw the numbers into the biggest computing clusters the world has ever seen and let statistical algorithms find patterns where science cannot."[1]

We can understand the enthusiasm here. AI is powerful stuff—and nobody knows for sure whether machines will someday be smart enough to design new medicines, tell us how the mind works, or invent a quantum theory of gravity, all from raw data alone.

But today? We're not even close. To illustrate why, we'll consider a simple, very specific scientific question: Do osteoporosis drugs cause cancer of the esophagus? This is exactly the kind of question that people who work on AI for health care would love to be able to answer automatically, using fancy algorithms turned loose on enormous databases of health information. In fact, this turns out to be a *perfect* question to look at, because some very smart people disagree about the answer. For example, Dr. Jane Green, a cancer epidemiologist at the University of Oxford, compiled evidence that osteoporosis drugs *do* cause cancer. Dr. Chris Cardwell, a public health researcher at Queen's University in Belfast, counters that they don't. Wouldn't it be great if we could use AI to help resolve the dispute?

First, some background. Many people with osteoporosis are prescribed drugs called "bisphosphonates."* These drugs can slow down or prevent bone loss, but they also carry a risk of upsetting your digestive tract, resulting in nausea or diarrhea. Some doctors worry that bisphosphonates might also increase your risk of developing esophageal, gastric, or colorectal cancer.

What does the evidence say? Let's start with the case for "no." Dr. Chris Cardwell and his research collaborators in Belfast examined a huge anonymized medical database with information on about 4 million patients in the United Kingdom. Their study design was simple. First, they looked for a group of patients in the database who'd used bisphosphonates. Then they ran a sophisticated matching algorithm to find "control" patients who were similar to the first group but hadn't used bisphosphonates. Finally, they tracked both groups through the database over time. Ultimately, they found no differences: bisphosphonate users and nonusers had similar rates of esophageal cancer. They published their findings in *JAMA*, the *Journal of the American Medical Association*, one of the world's most prestigious medical journals, in August of 2010.[2]

Now let's see the case for "yes." Dr. Jane Green and her research team at Oxford also examined a huge database of patients in the United Kingdom, and their study design, while different, was also simple. First they looked for "cases," or patients who'd developed esophageal cancer. Then they used a sophisticated matching algorithm to find "control"

* Pronounced "biss-FOS-fun-ates."

patients who were similar to the cases but hadn't developed cancer. Finally, they compared the cases with the controls, and they found that frequent bisphosphonate users had twice the risk of esophageal cancer as nonusers. They published their results in Britain's *BMJ*, another of the world's most prestigious medical journals, in September of 2010—just one month after the Cardwell article appeared in *JAMA*.[3]

So to summarize: one study says *no* extra risk, and the other says *double* the risk.[†] At least one of them must be wrong.

It's not unusual for two different published research findings to look at different data sets and find different answers to the same question, especially about something as complicated as human health. That's often how science works. At first some evidence points one way, and some points the other. Only over time does the evidence accumulate convincingly in a single direction.

Even so, there's something very striking about the Green and Cardwell studies, which were published a month apart on opposite sides of the Atlantic, and which reached opposite conclusions about whether bisphosphonates increase the risk of cancer. We've actually left out one very important part of the story. Without realizing it, both teams were running their analyses on the *same database*—specifically, the U.K.'s General Practice Research Database, which is publicly available to any health researcher. The teams got different answers, despite having the same cancer cases, the same bisphosphonate users, the same pool of control subjects . . . the same everything.

Well, not quite everything. The two studies got different answers because they made different *assumptions*. For example, Cardwell and team selected control patients on the basis of bisphosphonate exposure (a "retrospective cohort" design), while Green and team selected controls on the basis of cancer outcome (a "case-control" design). That's the biggest difference in assumptions between the studies, but it's far from the only one—and there's not a machine on the planet that could tell you which set of assumptions is right. That's because there's no

† An important point: twice a small number is still a small number. The baseline risk of esophageal cancer for people ages 60–79 is about 1 in 1,000 over five years. Green et al. estimated that this increases to about 2 in 1,000 with five years' use of bisphosphonates.

algorithm yet invented that can propose, test, and justify its own assumptions. Algorithms just do exactly what they're told.

Now you know why we're so skeptical of the AI evangelists on this issue. If a machine can't even tell you which bisphosphonate study is right after having seen their answers, then how could it possibly come up with the right answer on its own, without human help?

The lesson is simple. It may seem like we depend on smart machines for everything these days. But in reality, they depend on us a lot more.

A Study in Assumptions

In what ways does AI rely on assumptions made by people? What do these assumptions even look like? Why are they so important, and how do things go wrong when they're violated? These are the questions we'll address in this chapter.

In our view, the existence of clever AI doesn't somehow make assumptions less important. It makes assumptions *more* important, because the consequences of a single bad assumption can be amplified a millionfold or more, as some machine keeps repeating the same bad decision over and over again. Said another way: AI allows the fruit of a poison tree to grow exponentially larger. When this happens, it's usually because people have made poor choices in tending the soil.

There are three prime ways in which this can happen:

1. Rage to conclude.
2. Model rust.
3. Bias in, bias out.

To illustrate these themes, we'll ask for a little bit of help from a midcentury American icon: Joe DiMaggio.

Born in 1914 to a family of Italian immigrants in California, Joltin' Joe DiMaggio would go on to become one of the greatest baseball players ever, and a man whose fame transcended his sport. Ordinary people regarded him as a folk hero, and writers and artists—from Hemingway to Madonna, Rodgers and Hammerstein to Simon and Garfunkel—mentioned him in their most enduring works. An announcer at Yankee Stadium nicknamed him the "Yankee Clipper," after a new Pan American airliner; both were fast and glamorous.

As two probability nerds, we will remember DiMaggio most for the summer of 1941, when he got a hit in 56 straight baseball games. This record-smashing streak is still, at the time of writing, the longest ever—in fact, it towers over the second-place hitting streak of 45 games, by "Wee" Willie Keeler, in 1897. Most baseball fans consider DiMaggio's record unbeatable; Stephen Jay Gould, the eminent biologist and baseball fan, once called it "the most extraordinary thing that ever happened in American sports." As Gould put it, not only did DiMaggio successfully beat 56 Major League pitchers in a row, "he beat the hardest taskmaster of all . . . Lady Luck."[4]

Just how improbable was Joe DiMaggio's all-time record hitting streak of 56 games? The answer is certainly of interest to sports fans, who love to compare athletic feats across different eras and different sports—like whether DiMaggio's hitting streak was more impressive than Pelé's 1,281 career goals, or Michael Phelps's 23 Olympic gold medals.

We're actually interested in this question for a very different reason, though. DiMaggio's 56-game hitting streak can teach us a lesson about the importance of assumptions—specifically, about the dangers of using poor assumptions to extrapolate too far from the data. This lesson is fundamental to AI, because good data-science practices are essential for building machines that can learn and make decisions on their own. DiMaggio's hitting streak is the opening act in a parable about how the human side of this process can go wrong.

Joe DiMaggio and the Rage to Conclude

Act 1: The Streak

To calculate a probability for Joe DiMaggio's 56-game hitting streak, we'll begin with a metaphor. Suppose that a baseball game is like a coin flip: heads means DiMaggio got a hit in that game, and tails means he didn't. This metaphor makes it possible to analyze a hitting streak mathematically. We'll start with an easy one: What are the chances of getting heads twice in a row? For an actual coin, everyone would agree that the answer is ½ × ½ = ¼, since the coin comes up heads each time with probability ½, and since the first flip doesn't affect the second flip. It's only a little bit different for our hypothetical Joe DiMaggio coin: here the probability of heads is more like 80%, since he got a hit in about 80% of his games over the 1940–42 seasons.[‡] Therefore, the probability of a two-game hitting streak is 0.8 × 0.8 = 0.64.

This logic is easy to extend to longer streaks, using something called the "compounding rule." Suppose that some event happens with probability *P* in any one encounter. Then the chance it happens every time on *N* independent encounters is equal to P^N, or *P* multiplied by itself *N* times. So to calculate the probability of a 56-game Joe DiMaggio hitting streak, we multiply 0.8 by itself 56 times in a row. The result is a pretty small number:

P(56-game DiMaggio streak) = 0.8 × 0.8 × . . . × 0.8 = 1 in 250,000.

One natural reaction here is to think: Wow, wasn't Joe DiMaggio lucky? That's undeniably true. If you look at his streak game by game, you will undoubtedly find a few lucky bounces or weak hits that barely squeaked through.

But we marvel at DiMaggio's skill, not his luck. To see why, let's run through the same calculation using the statistics of a different player: Pete Rose, who had his own famous hitting streak in 1978. Around that time in his career, Rose was hitting safely in about 76% of his games. This was only 4% lower than DiMaggio's per-game hit probability of 80%. Yet over 56 games, the compounding rule magnifies this modest one-game difference into an enormous gulf of probability:

P(56-game Rose streak) = 0.76 × 0.76 × . . . × 0.76 = 1 in 5 million.

This is 20 times smaller than DiMaggio's figure of 1 in 250,000. And Rose himself was an extraordinary player. What about for an average

‡ We're aggregating data over three seasons to get a bigger sample size and to avoid cherry-picking DiMaggio's statistics from the games in his hitting streak, which would artificially inflate his actual per-game hit probability.

Major League player with his .250 batting average, who hits safely in about 68% of games?

$$P(\text{56-game streak}) = 0.68 \times 0.68 \times \ldots \times 0.68 = 1 \text{ in 2 billion.}$$

This will almost surely never happen.

So it is certainly true that DiMaggio needed a few friendly bounces during his streak. But he also needed to be very skillful in the first place, so that the odds he needed to overcome were "only" 250,000 to one.

Intermission: Model Versus Reality

We can learn two lessons about data science and AI from our analysis of Joe DiMaggio's hitting streak.

The first lesson is that probability compounds surprisingly fast, like the interest on a credit card. Over the long run, small edges become big edges. Think of how a small one-game difference in probability between DiMaggio (80%) and Rose (76%) compounded so dramatically, becoming a 20-fold difference over a 56-game span. In fact, this is a nice metaphor for how machines typically win when they play games against humans, whether the game is chess, Go, or recommending movies: they find many small advantages that compound together to form a big advantage.

The second lesson is about the importance of modeling assumptions. As you'll soon discover, if you learn the first lesson but not this second one, the result can be trouble.

Most calculations in data science require assumptions of one kind or another. We implicitly made two of them in analyzing DiMaggio's hitting streak. The first was *constant probability*: the chance that DiMaggio gets a hit is the same in every game (80%). The second assumption was *independence*: if DiMaggio gets a hit in one game, it tells us nothing about the next game. This is like saying that if you flip a coin twice, the first flip doesn't affect the second flip. Without these assumptions, the coin metaphor doesn't work, and neither do our calculations.

So are these assumptions literally true? Not exactly! Take the assumption of constant probability. Some games DiMaggio played at home, in a cavernous Yankee Stadium; others were on the road, in smaller ballparks. Some days there were fastballs; other days there were curve balls. Some days he faced Hall of Fame pitchers; other days he faced journeyman relievers barely out of the minor leagues. Far from the same probability every game, Joe DiMaggio had a different probability of getting a hit on every swing.

What about independence? That assumption is more debatable but probably still false. One recent study presented at the 2016 MIT Sloan Sports Analytics Conference examined a huge amount of historical baseball data and discovered clear evidence for a "hot-hand" effect among baseball hitters.[5] In other words, players who get a hit once are statistically more likely to get a hit the next time. This finding contradicts our assumption of independence.

So if our assumptions are wrong, you might ask, why even bother with the calculations in the first place? That's a great question, with a complicated answer.

Any scientist or engineer will tell you that models make the world go round. Boeing uses wind-tunnel models to help them build airplanes. Biologists use fruit flies as a model to help them understand human genetics. Toyota uses crash-test dummies as a model for what happens to people in a head-on collision. All these situations involve some assumptions about which features of the model have to be exact and which ones can be approximate. On many questions we couldn't make progress at all without models. As a lead engineer on the Mars Viking project once remarked, his job wasn't to design a probe that could land on Mars; it was to design a probe that could land on the model of Mars created by NASA's geologists.[6]

Data scientists use models, too, just like the one that helped us reason about Joe DiMaggio's hitting streak. Our models are based on probability. They're used to extract insights from data and to build successful AI systems, like many of the ones you've met throughout this book.

Data scientists have a favorite saying: all models are wrong, but some models are useful.[7] In other words, no model can describe the real world perfectly, but sometimes the mismatch is important, and sometimes it isn't. The corollary is that judging whether a model is useful requires knowing both the model *and* how that model will be used. A shop-window mannequin is a perfectly adequate model of a person if all you want to do is show off clothes, but it's a terrible model for training medical students about vascular anatomy.

So let's revisit our earlier statement, that Joe DiMaggio had a 1-in-250,000 chance of getting a hit 56 games in a row. This is *not* a statement about DiMaggio the man but about DiMaggio the model. This model makes assumptions, like constant probability and independence, that intentionally trade realism for simplicity.

It wouldn't be that hard to correct our model's worst problems. We could calculate separate probabilities for home and away games, or we could adjust the numbers based on the pitchers DiMaggio faced.[8] Machines make this kind of thing easy. Of course, to even ask the question, you have to understand the shortcomings of the model in the first place, and if all you need is a rough approximation for an armchair debate with your friends, these extra steps probably aren't worth it. You don't need to agree that the model is right, only that it's useful enough for the purpose at hand—that while enriching the model might be nice, doing so wouldn't bring much extra insight to a low-stakes discussion about hitting streaks in baseball.

The larger point here is that building models is a job fit only for people. A machine can make predictions based on the assumptions with which it's programmed, but only people can check those assumptions. A machine can fit a model, but only people can use that model to ask the right questions. A machine can churn through millions of data points per second, but only people can decide which data points are appropriate to use in the first place. Good data science requires people and machines working together, because the difference between the model and the reality isn't always such a casual matter as it is in a debate about baseball.

To illustrate this point, we'll now turn to act 2 of the Joe DiMaggio story. In this act, you'll see how a major newspaper pushed the DiMaggio model of winning streaks way too far, and ended up needlessly scaring millions of people as a result. This story has a very important lesson for understanding the role of assumptions in AI.

Act 2: How Effective Is Your Method?

The Egyptians of the Lower Kingdom used a mixture of honey and sodium carbonate. The Mesopotamians preferred acacia leaves and lint. The ancient Persians used elephant dung and cabbages; the Renaissance Europeans, lily root and silkworm gut.

In modern societies, we have it a bit easier. Most people choose condoms or the pill, or they get voluntarily and painlessly sterilized.

Birth control is at least as old as civilization; the big difference from ancient times is that our methods actually work well. Since the 1960s, when effective contraception became widely available, birth rates across the industrialized world have plummeted. Today, some experience with contraception is nearly universal among sexually active adults in rich countries.[9]

We recognize that, for many people, the choice of when to use contraception, and what method to use, cannot be reduced to a single variable.[10] But one important question for everyone is the chance of getting pregnant if you use a particular method. It was with this question in mind that, in 2014, *The New York Times* published an article entitled "How Likely Is It That Birth Control Could Let You Down?"[11] The authors of the article began from a simple premise: the more times you use any method of birth control, the more opportunities there are for it to fail. To put some numbers behind this idea, the authors of the article looked at published data on the 1-year efficacy of 15 popular contraceptive methods. They used this data—together with their own calculations, which we'll describe below—to create a slick interactive chart that purported to show each method's long-term failure rate, out to 10 years.

We have used the same published data to replicate the *Times*'s calculations,[12] using the same methodology employed by the authors of the article, for a subset of nine of these methods. Our calculations, which you can see in Figure 7.1, agree with theirs. Each panel shows a different contraceptive method. The vertical axis shows the *Times*'s estimate for the probability of getting pregnant at least once if you use that method over the long run.

If the numbers in this figure surprise you, you are not alone: the *Times* article shocked a whole lot of people. For example, the article

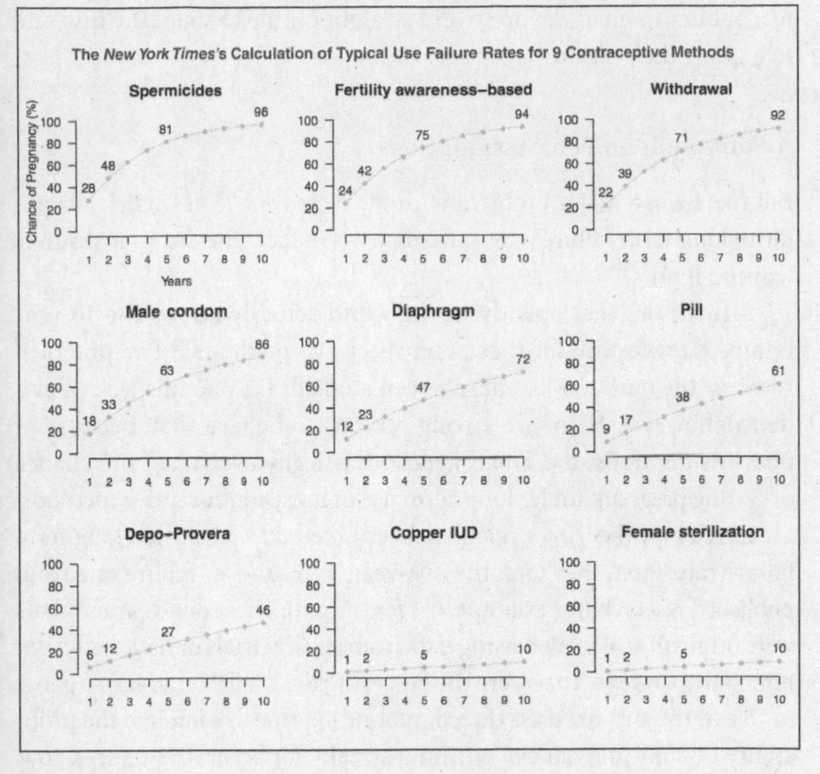

Figure 7.1.

stated that the 1-year failure rate for typical pill users was 9%,[§] but that the 10-year failure rate was an alarming 61%. The numbers for the condom were even worse: its 10-year failure rate was shown at 86%. To many people, these numbers seemed astonishingly high, and implied a far greater long-term risk of unplanned pregnancy than they were prepared for. Perhaps as a result, the article quickly went viral on social media—and while it may not have sparked a mass rush to join a convent, it did cause a lot of anxiety among ordinary *Times* readers, many of whom had presumably believed their own contraceptive methods to be more reliable. Even gynecologists, who should know the research

§ "Typical use" does not mean "correct use." If you use the pill exactly as instructed, the failure rate is much lower, at less than 1% per year.

just about as well as anyone, went straight online to share the link and to express their alarm.*

A Story Built on Poor Assumptions

But there was a major problem with the *New York Times* article: its putative long-term failure rates have no basis in fact. They're almost surely *way* too high.

It turns out that nobody in the world actually knows the 10-year failure rates for any of these contraceptive methods.[13] For practical reasons, the question just hasn't been studied. Despite this lack of evidence, however, there are strong reasons to believe that, because of poor assumptions, the *Times* article drastically overstated the chance of getting pregnant under long-term use of most contraceptive methods.

Here's how the *Times* calculated each method's purported long-term failure rate. First, they took the one-year "typical-use" failure rate from published research (for example, 9% for the pill). These one-year numbers were originally calculated using data from clinical trials or nationally representative surveys. They were the best estimates available. So far, so good.

Next, the authors used the compounding rule to calculate the probability of a no-pregnancy "winning streak" for several years in a row. Effectively, the *Times* journalists were treating a multiyear stretch of contraceptive use without a pregnancy exactly as if it were a Joe DiMaggio hitting streak, using the same two assumptions we used earlier: independence and constant probability across years.

Let's see an example. Among typical pill users, the probability of successfully avoiding pregnancy in year 1 was 91%. Based on this figure, the *Times* used the compounding rule to calculate the following probabilities:

$$P(\text{no pregnancy through 1 year}) = 0.91$$
$$P(\text{no pregnancy through 2 years}) = (0.91)^2 \approx 0.82$$
$$P(\text{no pregnancy through 3 years}) = (0.91)^3 \approx 0.75.$$

* For example, @hricciot: "Shocking—even to a gynecologist like me! #LARCisBest." "LARC" stands for "long-acting reversible contraception"—for example, an IUD.

And so on. By the time you get out to 10 years, the probability of a long "winning streak" starts to look pretty small: about 39%. This implies a 61% chance of at least one pregnancy over a 10-year period of typical pill use.

An Analogy

This calculation, however, has an enormous flaw. To see what it is, let's reason by analogy. Suppose that we conduct a study by recruiting 100 people and giving each of them a coin. These coins have been altered so that 90 of them have heads on both sides, and 10 of them have tails on both sides. Now we have our study participants start flipping their coins. We'll say that flipping tails is like getting pregnant. The question is: how many of our 100 study participants will go on a 10-year "no-pregnancy" streak, by successfully flipping heads 10 times in a row?

Clearly the answer is 90%: 90 out of 100 study participants have two-headed coins. They will avoid flipping tails forever. But let's see how we could get the wrong answer by using the compounding rule instead. Suppose we proceed as follows:

1. Take data from the first year of the study, in which 90 people flip heads and 10 flip tails.
2. Calculate the average probability of successfully avoiding tails in that first year, which will be 90%.
3. Use the compounding rule to calculate the probability of a ten-year winning streak based on the one-year estimate: 0.9^{10}, or about 35%.
4. Conclude that only 35 out of 100 study participants will successfully avoid pregnancy for 10 years in a row.

This is more or less exactly what the *Times* did in its analysis of contraceptive failure rates—and it's badly wrong. Is it correct to say that the average probability of flipping heads among study participants is 0.9? Absolutely. But does that mean that the average probability of flipping heads 10 times a row is 0.9^{10}, or 35%? Absolutely not. Ten people in our study will flip tails forever, and the other 90 people will flip heads forever. The population-average probability of a 10-flip winning streak—or a streak of any length—is actually 90%, not 35%. We can't even use the compounding rule as a rough approximation. The rule just doesn't work *at all* for population averages.

Here's a second analogy, one that's much closer to our question about contraceptive effectiveness: What are the chances that you can avoid causing a car accident for the next 10 years? Every year, there are 2 million drivers in the United States who cause an accident, which is 1% of the roughly 200 million drivers in the country. Thus the "typical" American's chance of making it through a single year without causing a car accident is about 99%. To calculate the probability of making it 10 years, you might be tempted to use the compounding rule, multiplying 0.99 by itself 10 times:

$$P(\text{10-year no-crash streak}) = 0.99 \times 0.99 \times \ldots \times 0.99 = 0.904.$$

But this is wrong. To understand why, let's rewind the clock back to the end of year 1. After the first year, the American population has cleaved into two groups: 2 million people who've caused a car crash, and 198 million people who haven't. Now ask yourself two simple questions. What will happen to each group's car insurance rates, and why?

The answer is clear. Group 1, with 2 million people who caused accidents, will see their rates go up. Group 2, with 198 million people who didn't cause accidents, will see their rates stay the same or go down. Why? The insurance companies aren't doing this to punish or reward people. They're doing it to price the risk of a *future* crash appropriately—and crashes in successive years aren't independent. Past crashes predict future crashes; some people are more likely to flip heads, and others are more likely to flip tails.

So what will happen in year 2? Almost surely, *more* than 1% of the people in group 1 will cause an accident in year 2. The drivers in this group are statistically less careful, at least on average. Similarly, *less* than 1% of the people in group 2 will cause an accident. The drivers in this group are statistically more careful—again, at least on average. In data science, we call this a lurking variable: something that has an important effect on the outcome of interest yet isn't directly measured.

The lurking-variable problem explains what was so wrong with our earlier calculation, where we took the average no-crash probability of 99% and compounded it out 10 years. The question is: Whose probability were we compounding? And the answer is: Nobody's! The annual probability of 1% is a property of a population—or at best, a property of some imaginary *Homo mediocritus* who has a 1% risk of a car crash, 2.1 children, half a college degree, one testicle, and one ovary. But every *real* person has a risk that's either higher or lower than the 1% average. If you crashed in year 1, your risk looks higher; if you didn't, your risk looks lower. The compounding-rule calculation is wrong for *literally everyone*.

Back to the Pill

Let's now return to the 10-year failure rate of the pill. Using a 1-year success probability of 91%, together with the compounding rule, the *Times* arrived at a 39% probability for a 10-year "winning streak" without a pregnancy. But as we've learned, you can't just compound up a population-average probability, because doing so doesn't account for lurking variables. And there's a really important lurking variable here: some people don't use a method the way they're supposed to, so they're more likely to "flip tails" and get pregnant early in the study. Other people are consistent users, so they're more likely to "flip heads" one year after another, avoiding pregnancy through the end of the study.

In fact, there really is no such thing as a typical *user* in a contraceptive study, only a typical *group*.[14] Contraceptive research isn't like some voyeuristic game of baseball, where scientists search the nation's bedrooms high and low for the average Major League player, who hits .250. It's about waiting and counting: you follow a typical group of people who are using a method—some of them erratically, some of them consistently—and you count how many of them get pregnant over time.

In any kind of situation like this, if you use the compounding rule to reason about what will happen to the group based on their 1-year average, you will get the wrong answer. To continue with our earlier analogies:

- In the first coin flip of our hypothetical coin-flipping study, 10% of people will flip tails. That figure includes both two-headed and two-tailed coins. So if you don't flip tails, we've learned something about your coin: it has two heads. Your chance of getting tails on the next flip is 0%.

- In any given year, about 1% of Americans will cause a car accident. That figure includes both good drivers and bad drivers. So if you don't cause a crash this year, we've learned something about your driving habits. Your chance of causing a crash next year is less than 1%.
- In their first year, about 9% of typical pill users will get pregnant. That figure includes both consistent and erratic users. So if you don't get pregnant this year, we've learned something about your pill-using habits. Your chance of getting pregnant next year is probably less than 9%. Maybe it's 8%, maybe it's 2%—nobody knows, because nobody's done the study. But we do know that the most erratic users, who contributed the most toward the 9% year-1 failure rate, have now dropped out.

Figure 7.2 conveys this idea. It compares the cumulative 10-year pregnancy rates among typical pill users, under two sets of assumptions. The dotted-line curve assumes what the *Times* assumed: that women who remain in the study in later years keep getting pregnant at the same unrealistically high rate of 9% per year. This predicts a cumulative failure rate of 61% over 10 years, with failures happening just as frequently in year 10 as in year 1.

Meanwhile, the solid black curve assumes that in later years, the women remaining in the study have *less* than a 9% average chance of getting pregnant, since the least-adherent users have already dropped out. This effect gets stronger over time, so that by the end of the study, only the most careful users remain. This curve predicts a cumulative pregnancy rate of more like 25% over 10 years, with the vast majority of contraceptive failures happening early in the 10-year window, and to erratic users.

We should emphasize that the only thing researchers actually know from the data is that 9% of pill users in a "typical use" cohort will get pregnant in year 1. From year 2 onward, both curves are extrapolations, based only on modeling assumptions.

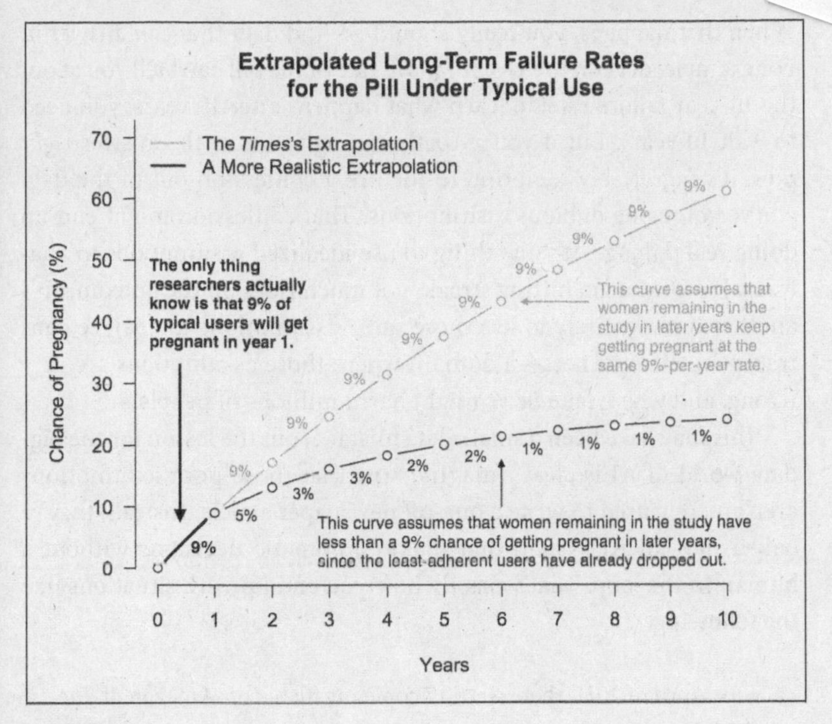

Figure 7.2.

But while all models are wrong, some models are more wrong than others.

Epilogue: A "Most Disastrous and Fruitless Mania"

We see the *Times* article on contraceptive failure rates as an example of what data-analysis guru Edward Tufte once called the "rage-to-conclude bias." He took the name from an aphorism of Flaubert's: "The rage for wanting to conclude is one of the most disastrous and fruitless manias to befall humanity."[15]

Tufte was referring to the human tendency to see patterns in randomness, but the rage-to-conclude phenomenon certainly doesn't stop there. Sometimes a data set is inherently unable to answer a question.

When that happens, you really should go find data that *can* answer it. For example, data on the 1-year failure rate of the pill can't tell you about the 10-year failure rate; to learn what happens after 10 years, you need to wait 10 years. But if you're really raging to know the answer *right now*, it's regrettably tempting to torture a confession out of the data you've got, using dubious assumptions. That confession might end up doing real damage. It's one thing to use idealized assumptions to analyze a Joe DiMaggio hitting streak; not much rides on the outcome. It's another thing entirely to use those same assumptions to analyze contraceptive effectiveness—a domain where those assumptions are *very* wrong, and where fake news might harm millions of people.

This may have been a small-data mistake, but the lesson for the big-data world of AI is clear. Imagine now that those poor assumptions aren't merely used to write a one-off newspaper article. Instead, they're baked into an AI system that makes automatic decisions without a human in the loop. That's exactly how you end up with situations like the following.

- In April of 2011, there were 17 copies available on Amazon of *The Making of a Fly*, a classic book about developmental biology. The cheapest of 15 used copies was $35.54, while the cheapest of two new copies was more than $23 million. It turned out that two algorithms, run by two different book sellers, had gotten into an inverse bidding war, under poor assumptions about the behavior of other sellers.[16]
- An online clothing retailer called Solid Gold Bomb created an algorithm that automatically made new designs for print-on-demand T-shirts, based on inserting random phrases into popular slogans, such as "Keep Calm and Carry On." Because of poor oversight, the company ended up accidentally advertising T-shirts emblazoned with terrible misogynistic phrases, including ones about sexual assault. It was a traumatic experience for many who encountered the designs online, and the company went out of business because of the backlash.[17]
- On May 6, 2010, U.S. stocks experienced a "Flash Crash," in which the market lost a trillion dollars of value in a matter of minutes—

> all because of algorithms gone wrong. According to the U.S. Department of Justice, a rogue trader based in London had submitted $200 million worth of "spoof" transactions that were modified 19,000 times over a very short period, before ultimately being withdrawn. This created a feigned sense of market pessimism about certain stocks. In response, everyone else's high-frequency trading algorithms—whose assumptions did not encompass the possibility of such feints—went completely haywire and issued millions of *real* "sell" orders. Before people figured out what was going on, the Dow Jones Industrial Average had lost 9% of its value within less than half an hour.[18] (Luckily, it rebounded almost immediately.)

These algorithms weren't aware of the consequences of their decisions or the business context that drove their creation in the first place. They were just doing what they were told to do, by people making poor assumptions.

Yet even as we recognize the danger of bad assumptions, it's also important to be measured in our skepticism, so that we aren't backed into an unhelpful corner where we find ourselves unwilling to make *any* assumptions, ever. Not all assumptions are bad, and not all the bad ones lead to problems. Artificial intelligence relies on pushing the frontiers of data science as far as possible, and sometimes that means relying on assumptions and approximations to extrapolate beyond the original domain of a data set. For example:

- Epidemiologists use AI to scan enormous databases of medical records to answer important health questions.
- Psychologists are studying Instagram posts to detect changes in someone's state of mind that might predict nascent depression or anxiety.
- Market watchers use social-media chatter as a leading indicator of economic activity.
- Zillow uses publicly available data, together with user-generated reports, to predict the market value of basically every house in the United States.

These ideas, and thousands more like them, can and do work. But to do so, they must circumvent a basic fact: most data sets in the internet age are collected under highly nonscientific conditions for someone else's purpose, and they are only incidentally useful for any other purpose. To get around that—or, just as importantly, to know when you can't—you had better reckon honestly with your assumptions.

A great democratization of AI is now under way. Ongoing efforts to collect and organize data, so that these powerful algorithms can do their job efficiently without making egregious mistakes, will create huge reservoirs of social and economic value. Soon almost every company, large and small, will rely on this kind of data to do business. In this new age, it's essential that we calm the rage to conclude, by remembering that every unverified assumption is a placeholder—an approximation to be used, for better or worse, only until more data is available.

Model Rust

You've now seen how poor assumptions, embedded into the very DNA of a model, can result in dreadful mistakes. Models aren't always born rotten, though. Sometimes they get that way from too much rust.

One especially famous and ill-fated application of AI illustrates this phenomenon perfectly. This system went online in 2008 in the hopes of addressing an important public health problem, with both money and lives hanging in the balance. Over time, however, its predictions drifted further and further out of line with reality. By 2012, the model was missing *badly.* Yet even as its performance deteriorated, the model continued to receive a huge amount of hype—after all, it used "big data," a buzzword so seductive, and yet so protected by a force field of mathematical tediousness, that it often deters scrutiny.

This system was called Google Flu Trends. This is the story of how it went wrong, and why it was finally taken offline in 2015.

Using Big Data to Predict Flu Outbreaks

Flu kills hundreds of thousands of people worldwide every year, and it is a source of misery for tens of millions more. And that's just seasonal flu. Infectious-disease experts lose far more sleep worrying about pandemic flu, such as the "Hong Kong flu" outbreak in 1968, or the "Spanish flu" outbreak in 1918, which killed 50 million people—three times as many as World War I.

To inform its flu-prevention and treatment efforts, the U.S. Centers for Disease Control and Prevention (or CDC) have long used something called ILINet: the Influenza-like Illness Surveillance Network. ILINet is a nationwide network of more than 2,700 health-care providers who send data and lab specimens directly to the CDC whenever they see patients with flu-like symptoms. The CDC uses that information to produce a weekly index of flu activity for every state.

Unfortunately, there's one big problem with ILInet: it might take a week or more for all the lab specimens to be processed and the raw data analyzed. So while public health agencies across the country depend on ILInet to make all kinds of important decisions about flu prevention, testing, and drug distribution—and while it's the best tool they have for situational awareness—it's usually two weeks out of date. That's enough time for a flu outbreak to infect a huge number of people.

Data scientists at Google, however, believed that they could solve this problem using a clever AI approach. Their insight was both simple and brilliant: the frequency of certain web-search queries ought to correlate strongly with flu activity. For example, Figure 7.3 shows how often Americans were asking Google "How long does flu last?" between 2008 and 2012.

Searches for this term peak every winter, right about when flu season does. Google can analyze these search queries much more quickly than the CDC can analyze lab samples from 2,700 clinics. It should therefore be possible to track flu activity with a much shorter reporting lag, using an AI system that aggregates search queries and produces a forecast.

The mapping between Google searches and flu activity, however, is imperfect and noisy. Not everyone uses the same search terms; nor does every search for a specific term mean that someone has the flu. You therefore wouldn't want to build a flu-tracking system naïvely, by counting one flu case for every Google search with the word "flu" in it. You have to be smarter than that, by building a prediction rule using historical data.

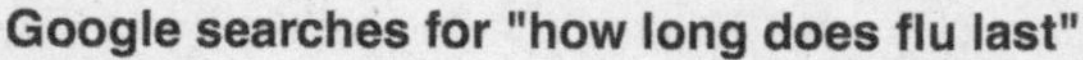

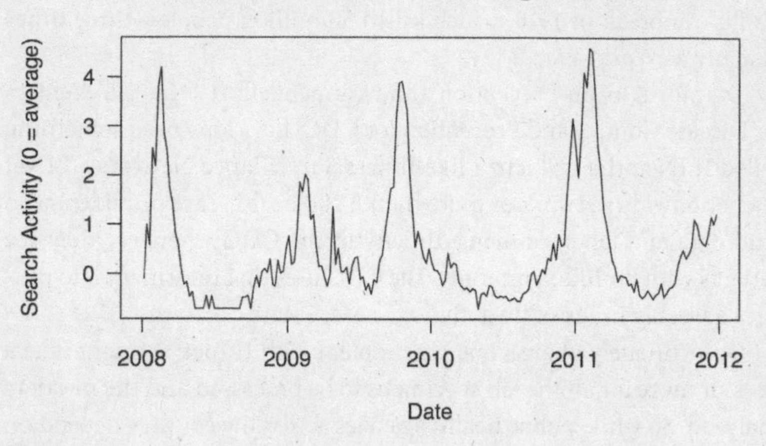

Figure 7.3.

Google's data scientists did exactly that. The inputs to their model were the frequencies of 50 million different possible search terms. The output was a prediction for the published ILInet figures provided each week by the CDC—the gold standard for quantifying flu activity in the United States. The Google team described their method in a paper in *Nature*,[19] and they began posting their model's flu-activity forecasts on a dedicated Flu Trends website, to great fanfare from the public health community.

Unfortunately, Flu Trends had a rocky start. In 2009, it completely missed a huge out-of-season spike in flu caused by the H1N1 ("swine flu") pandemic. In response, Google's engineers made some tweaks to the algorithm, and for the next two years, from the autumn of 2009 to the summer of 2011, Flu Trends performed fine: it tracked the CDC's numbers pretty closely, but without the two-week lag.[20]

Starting in autumn 2011, however, it all went south. In the 2011–12 season, the model overestimated flu activity by about 50%, raising alarm bells among public health professionals who'd come to trust it. Then it got worse: in the 2012–13 season, the Flu Trends prediction overshot the winter peak by almost 150%.[21] Overall, from August 2011 to September 2013, the Google estimates were too high in 100 out of 108 weeks.[22] If public health officials had been relying on these estimates, they would have ended up devoting resources to handle tens of thousands of flu cases that simply didn't exist.

How Models Age

Google Flu Trends is a great example of a general principle: in AI, models don't stay factory-fresh for long.

In fact, we like to think of models aging like a cast-iron skillet. If you look after an iron skillet well, it actually gets better with time: a nice patina builds up, and food won't stick as easily. The same is true of a model in AI. If you maintain a model and regularly "season it" with new data—that's the trial-and-error model-fitting process we talked about in chapter 2—it makes better predictions over time. But if you neglect the model—if you let a crust of old, blackened assumptions build up too far—then patina can easily turn to rust. Neglect it some more, and eventually the model just rots away into nothing.

Flu Trends suffered from a serious case of model rust, verging on rot. To understand why, we spoke to Dr. Rosalind Eggo, an infectious-disease researcher at the London School of Hygiene & Tropical Medicine. She was quick to commend Google for having put such a rich data resource to use in the service of a public good, but she also thought that the manner in which Flu Trends missed the 2009 H1N1 pandemic should have raised a red flag. "Although Google was very opaque about the details of the algorithm," Eggo explained, "some people conjecture that the search terms that got picked up weren't flu terms at all, but rather were *winter* terms, like searches for high-school basketball. As a result, they were just capturing a general seasonality in search patterns, rather than something specific to flu." Eggo quoted from a 2014 *Science* paper, by Dr. David Lazer and colleagues, that examined Flu Trends' performance over time and concluded that the original algorithm was "part flu detector, part winter detector."[23] This should have made people a bit more suspicious of the design choices that went into the algorithm.

Another big issue was that Google actually nudged its own users to violate the modeling assumptions of Flu Trends. Google is constantly tweaking how its search algorithms work in thousands of ways both large and small—and people tweak their own search patterns in response. One example is Google's autocomplete feature, which suggests search terms as you type. This saves people time, but it also changes their behavior. Eggo points out that when people put snotty fingers to keyboards to get advice about their flu symptoms, "auto-complete will affect what they end up searching for." For example, maybe somebody who'd previously searched for "best flu medicine" now searched for "best flu treatment" instead, because "treatment" was the top autocomplete suggestion. But Flu Trends relied on the assumption of a stable relationship between search terms and flu activity. If that assumption breaks down—if something *other than flu* causes big changes in what terms people are searching for—then the forecasting model breaks down, too. That's probably what happened post-2009: these thousands of tiny business-driven changes in the algorithms, according to Eggo, "were not being tracked by Google Flu Trends, and their effect on the quality of the fit was not being monitored."

There are two sad aspects of this story. First, in principle, it should not have been hard for a company with the resources of Google to make an ongoing commitment to "rust prevention," by allowing its model to adapt to new search behaviors. "Dynamic" models, the kind that can track a changing set of structural relationships between inputs and outputs, are a standard part of the data-science toolkit. It's something of a mystery to us why Google's data scientists didn't take things in this direction—and as far as we're aware, no one there has ever explained why in public.

Another sad fact is that many researchers have now been deterred from pursuing an idea with such wonderful potential. We asked Eggo whether the public health community had learned its lesson in the wake of Flu Trends, and she replied:

> I think they've learned a bit too much of a lesson. They've been frightened by this failure. If Google simply allowed their model to adapt with the search algorithm, it would probably work fine. And it could

> give you much more detailed city-level information than the formal surveillance systems could ever hope to provide. There's so much promise in that. But I'm guessing that Google didn't want any more bad press. And to the researchers, it's a case of once bitten, twice shy.

Sometimes all you need is better rust prevention. We hope that Google is willing to try again.

Bias In, Bias Out

Another big issue in AI arises when we train our models using a data set with an inherent bias. Here's an analogy. In the 2016 presidential election, poll aggregators all predicted a Hillary Clinton victory. But their forecasting models, no matter how clever, were inherently limited by the quality of their input data. The trouble was a small but persistent bias in the underlying polls, which underestimated support for Donald Trump.

Many algorithms in AI suffer from a similar problem: bias in, bias out. There's a classic parable here about a neural-network model the U.S. Army once built to detect tanks that were partially hidden on the edge of a forest.[24] Army scientists trained their model using a labeled data set of photographs, some with tanks and some without. The neural network turned out to have surprisingly high accuracy. It even did well when the army held out some of the original training data and used it exclusively to test the performance of the model. (This practice of validating results using notionally "out-of-sample" data is standard in AI.)

But when the army tried to use the model to detect tanks in the real world, it didn't work—you might as well have been flipping a coin. The experts were puzzled. What could explain the dramatic drop-off in performance? Then someone realized that the training data had a hidden bias. All the photos with tanks had been taken on a sunny day. All the photos without tanks had been taken on a cloudy day. What the model had actually learned to do was to distinguish a forest with and without shadows cast by trees—a skill that was completely useless for identifying tanks.

Many unwise or unfair applications of AI fail for a similar reason:

hidden biases in the training data. The more data you have, the worse this problem can get. Bigger data sets don't necessarily eliminate bias. Sometimes, they just asymptotically zero in on the biases that have always been there.

Take, for example, the manner in which artificial intelligence has been used in the criminal justice system. Judges responsible for sentencing decisions have always tried to assess the danger that a convict poses to society. Traditionally, they've done this in an unscientific way, by drawing on their own knowledge, intuition, and experience to make judgments about a defendant's character and record. Now, however, those assessments are starting to be informed by data—and some judges today are even relying on machine-learning algorithms, trained on historical data from the justice system, that predict someone's likelihood of recidivism.

One popular recidivism-prediction algorithm is called COMPAS, for "Correctional Offender Management Profiling for Alternative Sanctions." COMPAS, like all such systems, is explicitly prevented from "knowing" about things like a defendant's race or gender as one of its inputs. But that's not enough to prevent bias from creeping in: the whole premise of machine learning is that it's possible to learn by proxy about unobserved attributes. So to test the neutrality of the algorithm, journalists at ProPublica looked into the recidivism-prediction scores for 10,000 people arrested in Broward County, Florida, and sentenced by judges using COMPAS.[25] They checked to see who had been arrested again within two years, and they found a striking racial disparity. Among people who did not commit further crimes, black defendants had a higher false-positive rate: they were more likely than white defendants to have been wrongly classified as high risk. Conversely, among people who *did* commit further crimes, whites had a higher false-negative rate: they were more likely than blacks to have been wrongly classified as low risk.

How could this have happened if the underlying algorithm were truly "race neutral"? One very plausible explanation for the discrepancy is the *data itself*. Remember, AI algorithms are designed to find and recreate the patterns in the data sets they were trained on. If those patterns are inherently discriminatory, then an algorithm will learn to discriminate. Suppose you accept the arguments, as some scholars have put forward, that police are more likely to arrest a black person for the same crime; that prosecutors are more likely to pursue cases involving black suspects; that juries are more likely to convict black defendants; and that whites are more likely to have better lawyers. If any of those claims are true, then *of course* the data would reflect higher recidivism rates among blacks than whites, for reasons having nothing to do with anyone's propensity to reoffend. If dark skin predicts that someone is more likely to be caught, prosecuted, and convicted for a given crime, then *any* recidivism-prediction algorithm doing its job will try its hardest to find proxies for dark skin. And given the long, sad history of racial inequality in the United States, those proxies are abundant; the algorithm might ask, for example, whether the defendant has any family members in jail.

Unfortunately, we can't ascertain whether this kind of racism-by-proxy was responsible for the pattern of bias in the COMPAS algorithm's predictions in Broward County, Florida. That's because the algorithm is secret.[26] The company that sells the software won't tell defendants or judges about its inner workings—so if the algorithm classifies you as high risk, and the judge gives you a longer sentence as a result, *you can't even ask why.* This is morally obscene. As a culture, we don't accept secrecy in the algorithms used to rank college football teams. Why should we accept it in a situation where someone's freedom is on the line?

Many people, when they hear about something as shocking as a secret algorithm handing down prison sentences in a racially biased way, reach a simple conclusion: that artificial intelligence should play no role whatsoever in the criminal justice system.

While we're as shocked and angry as anyone, we think that's the wrong conclusion. Yes, we must all fight algorithmic bias when it arises. To do that, we need constant vigilance by experts: people who know the law but who also know AI, and who are empowered to act if they see a threat to justice. But even as we acknowledge the pitfalls of using AI to help people make important decisions, and even as we echo the call for transparency and fairness to become defining values of this new age, let's not forget that there's also *incredible* potential here. After all, it is *people,* rather than algorithms, who are primarily responsible for the mess in Broward County:

- Those in the justice system who've treated a sentencing algorithm in the same way you would treat a microwave oven, by just punching in some numbers and walking away.
- The legislators and higher courts who have allowed these decisions to be made using proprietary algorithms whose internal workings cannot be interpreted, appealed, or even examined.
- Above all, the police, prosecutors, judges, and juries whose actions collectively encoded a *human* racial bias in the data sets on which these algorithms are trained.

It's this last group—which includes every single one of us—that should worry you the most. If you hear the story of COMPAS and jump to the conclusion that AI should be kept miles away from important decisions, we'd ask you a simple question: Is the status quo really your friend here? Important decisions in the criminal justice system have *always* been made using biased algorithms trained on flawed data. It's just that those algorithms happen to live inside people's minds. You can't subject the biases of these "human wetware" algorithms to direct numerical scrutiny, like you can with a prediction rule in AI. But all you need to do is look at the roster of criminals on death row in Huntsville, Texas, or examine incarceration rates in America stratified by race—0.45% for whites, 2.31% for blacks—and you can see the damage those human biases have wrought.[27]

And it's not just in the criminal justice system. How would you like, for example, to leave your future in the hands of these decision-makers?

- The human-resources manager who's more likely to interview people with stereotypically white names than with stereotypically black ones.
- The boss who gives higher job evaluations to attractive employees.
- The college admissions director who holds Asians to a higher standard than whites.

- The executive who pays a woman 80 cents on the dollar to do the same job as a man.
- The diverse, well-intentioned, and basically decent people on the hiring committee who have to look at a large stack of résumés for a single job, and who are therefore unduly influenced by elegant typesetting and active verbs.

Biased and ill-informed decision-making algorithms are no less pernicious just because they run on little gray cells rather than little silicon chips. Wouldn't the world be a better place if people who suffer at the hands of prejudiced decision-makers actually had the recourse of a second opinion from AI—an algorithm whose reasoning and biases lived in the open, and could therefore be corrected?

Postscript

Imagine that you could travel back in time to the 1990s, when you downloaded your first web browser, or to the 2000s, when you bought your first smartphone and opened accounts on Facebook and Twitter. In light of what you've learned since then, what advice might you give yourself—about what information to share, what photos to post, and what habits to cultivate? Or if you had the ear of corporate chiefs and government regulators, what would you want them to know? What stories would you tell about how these technologies have changed your life for the better? What pathologies would you ask them to prevent?

Artificial intelligence will soon play a role in decisions that are much more important than the films people see on Netflix, the music they hear on Spotify, or the news stories they're recommended by Facebook. They will inform what medical treatments people receive, what jobs they compete for, what colleges they attend, what loans they qualify for—and yes, what jail sentences they receive when they commit a crime. In thinking about these complex issues, we can't rely on advice from a time traveler. It's just us, and *we have to get this right.* There is much to gain but also much to lose, and the balance we strike between costs and benefits will be affected enormously by whether the people in charge understand how AI technologies really work. If we muddle through—or worse still, if we let the world's tech companies just move fast and break stuff while the rest of us waste our time worrying pointlessly about sci-fi nightmares—we will kill the credibility of these AI systems before they even have a chance to mature, and we'll deprive humanity of so much promise.

But now imagine a world where we're actually smart about our efforts—a world where we put the right experts and the right legal protections in the right places, and where we're eternally vigilant about the biases and assumptions of our algorithms. In that world, our decision-making protocols could become *radically* better than the bias-riddled ones we have now—the ones that give an undeserved leg up to those with the prettier face, or the livelier manner, or the richer dad, or the whiter skin. Our collective vision and technology have reached a point where we can successfully teach machines to drive a car, predict kidney disease, and carry on a conversation. We can certainly teach those machines to play fair. They might even teach us.

Everyone agrees that some matters are too important to be settled by an unaccountable algorithm, operating alone. Some of us would just go a step further and say the same thing about people. When it comes to the important decisions in life, we can and should combine artificial intelligence with human insight and human values. All it takes is people and machines working together.

ACKNOWLEDGMENTS

Together we want to thank the two people most responsible for nurturing this book from its earliest stages: Lisa Gallagher of DeFiore & Company and Tim Bartlett of St. Martin's Press. This is the first thing either of us have written that wasn't for an academic audience, and we started with almost no sense of what writing and publishing a "real" book would actually entail. We are so grateful to Lisa for seeing and cultivating the potential in those first scribbled drafts, which now look so painfully clunky. We are equally grateful to Tim, both for taking a gamble on two data scientists foolhardy enough to try their hands at writing, and for giving us such unerring advice along the way. We are also indebted to Doug Young of Transworld for his valuable editorial feedback.

We also thank the many other people at DeFiore, St. Martin's Press, and Macmillan who have been so helpful, including Robert Allen, Alan Bradshaw, Jeff Capshew, Laura Clark, Jennifer Enderlin, Tracey Guest, Leah Johanson, Linda Kaplan, Alice Pfeifer, Gabrielle Piraino, Jason Prince, Sally Richardson, Brisa Robinson, Mary Beth Roche, Robert Van Kolken, Laura Wilson, and George Witte. We give special thanks to India Cooper, whose magnificent editing efforts have put into stark relief the difference between a professional writer and two amateurs like us. Thanks also to Larry Finlay, Bill Scott-Kerr, and the rest of the Transworld team for their support.

We thank Ellen Zippi for her invaluable help in researching this book. We are also grateful to many of our colleagues for sharing stories and expertise, most especially Steven Levitt for introducing us to Lisa Gallagher, and David Madigan for drawing our attention to the two studies on bisphosphonate usage described in chapter 7. We thank Rosalind Eggo, Katherine Heller, and Mark Sendak for their time and trouble in agreeing to be interviewed. Thanks also to those family members who tirelessly read early drafts and gave their feedback: Catherine Aiken, Patricia and Josh Lowry, Anne and George Scott, and Brian Woods.

PERSONAL ACKNOWLEDGMENTS

I am grateful to my co-author, James Scott. Above all, I thank my family for their love and support: my wife, Anne Gron, and our children, Emma, Michael, and Sarah.

—*Nick Polson*

Thank you to Nick Polson. I owe Nick for so much in my career that I cannot possibly list it all here; this book is but the latest in a long string of projects and ideas that he has so generously shared with me. I expect that over the decades to come, I will look back and see Nick as the single most important influence on my professional life, and the best friend I ever had in this field. I also want to thank the three most important teachers I ever had: Bill Jeffreys, Jim Berger, and John Trimble. Without Bill and Jim, I would never have become a statistician. Without John's kindness and generosity, I would never have known how to "tighten/sharpen/brighten" my way to better prose. I also thank my parents, who gave me so much—not least of which was their example. Finally, I am grateful to my wife, Abigail Aiken, for just about everything. I love you, and I could not have helped write this book without your support.

—*James Scott*

NOTES

CHAPTER 1

1. Quotation from Kevin Spacey in the James MacTaggart Memorial Lecture at the Edinburgh Fringe Festival, 2013. Video available on YouTube at https://www.youtube.com/watch?v=oheDqofa5NM.
2. Nancy Hass, "And the Award for the Next HBO Goes to . . . ," *GQ*, January 29, 2013. https://www.gq.com/story/netflix-founder-reed-hastings-house-of-cards-arrested-development.
3. Numbers taken from U.S. Census Bureau, *Statistical Abstract of the United States* (Washington, D.C.: USGPO, 1944, 1947, 1950), and Army Air Forces, *Statistical Digest (World War II)*, available at https://archive.org/details/ArmyAirForcesStatisticalDigestWorldWarII.
4. Material on the life of Abraham Wald was drawn from the following sources: W. Allen Wallis, "The Statistical Research Group, 1942–1945," *Journal of the American Statistical Association* 75, no. 370 (June 1980): 320–30; Marc Mangel and Francisco J. Samaniego, "Abraham Wald's Work on Aircraft Survivability," *Journal of the American Statistical Association* 79, no. 386 (June 1984): 259–67, and see also "Comment" by James O. Berger (267–69) and "Rejoinder" by the authors (270–71); J. Wolfowitz, "Abraham Wald, 1902–1950," *Annals of Mathematical Statistics* 23, no. 1 (1952): 1–13; Oskar Morgenstern, "Abraham Wald, 1902–1950," *Econometrica* 19, no. 4 (Oct. 1951): 361–67; Karl Menger, "The Formative Years of Abraham Wald and His Work in Geometry," *Annals of Mathematical Statistics* 23, no. 1 (1952): 14–20; L. Weiss, "Wald, Abraham," in

Leading Personalities in Statistical Sciences: From the Seventeenth Century to the Present, ed. Norman L. Johnson and Samuel Kotz (New York: John Wiley & Sons, 1997), 164–67; "Abraham Wald," *MacTutor History of Mathematics*, http://www-history.mcs.st-andrews.ac.uk/Biographies/Wald.html.

5. W. Allen Wallis, "The Statistical Research Group, 1942–1945," *Journal of the American Statistical Association* 75, no. 370 (June 1980): 320–30.
6. Our presentation of Wald's approach uses modern notation and terms and is therefore intentionally anachronistic. Wald did not frame the problem in precisely these terms. We have also left out a lot of technical detail. We encourage the interested reader to consult Marc Mangel and Francisco J. Samaniego, "Abraham Wald's Work on Aircraft Survivability," *Journal of the American Statistical Association* 79, no. 386 (June 1984): 259–67; see also "Comment" by James O. Berger, 267–69, and "Rejoinder" by the authors, 270–71.
7. W. Allen Wallis, "The Statistical Research Group, 1942–1945," *Journal of the American Statistical Association* 75, no. 370 (June 1980): 320–30; Mangel and Samaniego, "Rejoinder."
8. Available at https://www.netflixprize.com/community/topic_1537.html.
9. Dan Keating, Kevin Schaul, and Leslie Shapiro, "The Facebook Ads Russians Targeted at Different Groups," *Washington Post*, November 1, 2017, https://www.washingtonpost.com/graphics/2017/business/russian-ads-facebook-targeting/.
10. National Cancer Institute, "Study Shows Promise of Precision Medicine for Most Common Type of Lymphoma," July 20, 2015, https://www.cancer.gov/news-events/press-releases/2015/ibrutinib-lymphoma-subtype.

CHAPTER 2

1. Ihsan Hafez, "Abd al-Rahman al-Sufi and His Book of the Fixed Stars: A Journey of Re-discovery," PhD diss., James Cook University, 2010.
2. Marcia Bartusiak, *The Day We Found the Universe* (New York: Vintage Books, 2010), 52.
3. Agnes Clerke in *The System of the Stars*, 1890, in ibid., 53.
4. K. C. Freeman, "Slipher and the Nature of the Nebulae," Astronomical Society of the Pacific Conference Series, vol. 471 (2013), http://arxiv.org/abs/1301.7509.
5. Our two main references on Henrietta Leavitt were Nina Byers and Gary Williams, *Out of the Shadows: Contributions of Twentieth-Century Women to Physics* (Cambridge: Cambridge University Press, 2006), and Marcia Bartusiak, *The Day We Found the Universe* (New York: Vintage Books, 2010).
6. Shapley actually estimated that the Milky Way was 300,000 light-years across. Subsequent refinements have produced an estimate of 100,000 light-years across.
7. Quoted in Bartusiak, *The Day We Found the Universe*, 218.

8. Clara Moskowitz, "Star That Changed the Universe Shines in Hubble Photo," Space.com, May 23, 2011, https://www.space.com/11761-historic-star-variable-hubble-telescope-photo-aas218.html.
9. There is a priority dispute over least squares. Gauss, the great German mathematician, may have invented it first, although Legendre had the first published statement of the method. Those interested in the history should consult Stephen Stigler, "Gauss and the Invention of Least Squares," *Annals of Statistics* 9, no. 3 (1981): 465–74.
10. "John Deere Green," written by Dennis Linde, performed by Joe Diffie.
11. John Mannes, "This Beekeeper Is Rescuing Honeybees with Deep Learning and an iPhone," *TechCrunch*, May 2, 2017, https://techcrunch.com/2017/05/02/beekeepers/.
12. Alex Brokaw, "This Startup Uses Machine Learning and Satellite Imagery to Predict Crop Yields," *The Verge*, August 4, 2016, https://www.theverge.com/2016/8/4/12369494/descartes-artificial-intelligence-crop-predictions-usda.
13. Sam Shead, "Google's DeepMind Wants to Cut 10% Off the Entire UK's Energy Bill," *Business Insider*, March 13, 2017, http://www.businessinsider.com/google-deepmind-wants-to-cut-ten-percent-off-entire-uk-energy-bill-using-artificial-intelligence-2017-3.
14. "The Women Missing from the Silver Screen and the Technology Used to Find Them," *Google.com*, https://www.google.com/intl/en/about/main/gender-equality-films/.

CHAPTER 3

1. Statistics from the Insurance Institute for Highway Safety, http://www.rmiia.org/auto/teens/Teen_Driving_Statistics.asp.
2. "Claude E. Shannon: A Goliath Amongst Giants," https://www.bell-labs.com/claude-shannon/.
3. Les Earnest, "Stanford Cart," December 2012, https://web.stanford.edu/~learnest/cart.htm.
4. Thuy Ong, "Dubai Starts Testing Crewless Two-Person 'Flying Taxis,'" *The Verge*, September 26, 2017, https://www.theverge.com/2017/9/26/16365614/dubai-testing-uncrewed-two-person-flying-taxis-volocopter; Tom Simonite, "Mining 24 Hours a Day with Robots," *MIT Technology Review*, December 2016, https://www.technologyreview.com/s/603170/mining-24-hours-a-day-with-robots/; "Asia's First Automated Container Terminal, at Port of Qingdao, China," live report on New China TV, May 11, 2017, https://www.youtube.com/watch?v=bn2GPNJmR7A.
5. Peter Henderson, "U.S. Judge Deals Setback to Waymo Damage Claim in Uber Lawsuit," Reuters, November 3, 2017, https://www.reuters.com/article/us

-alphabet-uber-lawsuit/u-s-judge-deals-setback-to-waymo-damage-claim-in-uber-lawsuit-idUSKBN1D32J0.

6. William Beecher, "Vast Search Fails to Find Submarine," *New York Times*, May 29, 1968, A1.
7. "The President's News Conference of May 28, 1968," in *Public Papers of the Presidents of the United States: Lyndon B. Johnson, 1968–1969* (Washington, D.C.: USGPO, 1970), 656.
8. This and subsequent details about the Palomares incident are drawn from Sharon Bertsch McGrayne, *The Theory That Would Not Die* (New Haven: Yale University Press, 2011), 182–94.
9. Ibid., 192–94.
10. PBS *Nova* documentary, "Submarines, Secrets, and Spies," originally broadcast January 19, 1999, available at https://www.youtube.com/watch?v=RvJTAMQQQUY.
11. McGrayne, *The Theory That Would Not Die*, 202.
12. PBS *Nova* documentary, "Submarines, Secrets, and Spies."
13. McGrayne, *The Theory That Would Not Die*, 202.
14. David M. Eddy, "Probabilistic Reasoning in Clinical Medicine: Problems and Opportunities," in *Judgment Under Uncertainty: Heuristics and Biases*, ed. Daniel Kahneman, Paul Slovic, and Amos Tversky (Cambridge: Cambridge University Press, 1982), 249–67.
15. It's actually 99 false positives, but we're rounding off to 100 to keep the numbers easier to work with. If you correct for our modest round-off error, the actual posterior probability *P*(cancer | positive test) is really 7.5%, not 7.4%.
16. Madison Marriage, "86% of Active Equity Funds Underperform," *Financial Times*, March 20, 2016, https://www.ft.com/content/e555d83a-ed28-11e5-888e-2eadd5fbc4a4.
17. See the technical sidebar on page 106.
18. Lawrence D. Stone, *The Theory of Optimal Search* (New York: Academic Press, 1975).
19. Lawrence D. Stone, Colleen M. Keller, Thomas M. Kratzke, and Johan P. Strumpfer, "Search for the Wreckage of Air France Flight AF 447," *Statistical Science* 29, no. 1 (2014): 69–80.
20. "Breakthrough: Robotic Limbs Moved by the Mind," *60 Minutes*, originally broadcast December 30, 2012, available at https://www.cbsnews.com/news/breakthrough-robotic-limbs-moved-by-the-mind-30-12-2012/.

CHAPTER 4

1. A. Bartoli, A. De Lorenzo, E. Medvet, and F. Tarlao, "Your Paper Has Been Accepted, Rejected, or Whatever: Automatic Generation of Scientific Paper

Reviews," in *Availability, Reliability, and Security in Information Systems*, ed. F. Buccafurri et al. (New York: Springer Berlin Heidelberg, 2016), 19–28.

2. Andy Pandy, January 18, 2016, https://twitter.com/_Pandy/status/689209034143084547.
3. A minor technical point: for our purposes, there is no need to distinguish between a "compiler" (for a language like C++ or Java) and an "interpreter" (for a language like Python). We use the term "compiler" to encompass both concepts here.
4. Kathleen Broome Williams, *Grace Hopper: Admiral of the Cyber Sea* (Annapolis, Md.: Naval Institute Press, 2004), 1.
5. Ibid., 2.
6. Ibid., 11.
7. Ibid., 18–20.
8. Ibid., 22.
9. Ibid., 26.
10. Ibid., 29.
11. Ibid., 27–28.
12. Ibid., 82.
13. Kurt W. Beyer, *Grace Hopper and the Invention of the Information Age* (Cambridge, Mass.: MIT Press, 2009), 53.
14. Douglas Hofstadter, *Gödel, Escher, Bach: An Eternal Golden Braid* (New York: Vintage, 1980), 290.
15. Williams, *Grace Hopper*, 70.
16. Ibid., 80.
17. Ibid., 85.
18. Ibid., 86.
19. Ibid.
20. See ibid., 87. Original reference in Richard L. Wexelblat, ed., *History of Programming Languages I* (New York: ACM, 1978), 17.
21. Bruce T. Lowerre, "The HARPY Speech Recognition System," Ph.D. thesis, Department of Computer Science, Carnegie Mellon University, 1976.
22. "10 Inexplicable Google Translate Fails," https://www.searchenginepeople.com/blog/10-google-translate-fails.html.
23. For details of the method as well as extensive accuracy evaluations, see Yonghui Wu et al., "Google's Neural Machine Translation System: Bridging the Gap Between Human and Machine Translation," October 8, 2016, https://arxiv.org/abs/1609.08144.
24. Peter Norvig, "On Chomsky and the Two Cultures of Statistical Learning," http://norvig.com/chomsky.html.
25. If you want to get really technical, these "probe words" are really called "context vectors." See Tomas Mikolov et al., "Distributed Representations of Words

and Phrases and Their Compositionality," *Advances in Neural Information Processing Systems* 26 (NIPS, 2013), https://papers.nips.cc/paper/5021-distributed-representations-of-words-and-phrases-and-their-compositionality.

26. Tomas Mikolov, Wen-tau Yih, and Geoffrey Zweig, "Linguistic Regularities in Continuous Space Word Representations," in *Proceedings of NAACL-HLT, 2013* (Stroudsburg, PA: Association for Computational Linguistics, 2013), 746–51.

CHAPTER 5

1. We distinctly remember hearing this piece of commentary on a TV show in the wake of the coin-flip incident, but we have been unable to find a transcript of the show online. Our apologies to the witty and sadly anonymous commentator.
2. Stephen Quinn, "Gold, Silver, and the Glorious Revolution: Arbitrage Between Bills of Exchange and Bullion," *The Economic History Review* 49, no. 3 (1996): 479–90.
3. Thomas Levenson, *Newton and the Counterfeiter* (Boston: Mariner Books, 2010), 626–63.
4. John Craig, *Newton at the Mint* (Cambridge: Cambridge University Press, 1946), 6–7.
5. Ming-hsun Li, *The Great Recoinage of 1696 to 1699* (London: Weidenfeld and Nicolson, 1963), 47.
6. Levenson, *Newton and the Counterfeiter,* 137–38.
7. Thomas Babington Macaulay, *The History of England from the Accession of James II*, Volume 1 (New York: Harper & Brothers, 1856), 187.
8. Details of the inspection process used in the Trial of the Pyx are described in: Stephen M. Stigler, *Statistics on the Table* (Cambridge, Mass.: Harvard University Press, 1999), 386–89.
9. Ibid., 389–90.
10. John Craig, *The Mint: A History of the London Mint from A.D. 287 to 1948* (Cambridge: Cambridge University Press, 2011), 212, emphasis added.
11. Ibid., 104.
12. Stigler, *Statistics on the Table*, 391.
13. Craig, *Newton at the Mint*, 12–14.
14. Levenson, *Newton and the Counterfeiter,* 139–41.
15. Ibid., 141–44.
16. The Great Recoinage is judged by economic historians to have been a failure as monetary policy. The point we're making here is simply that it was successful as an industrial undertaking, quite apart from its economic effects.
17. Craig, *Newton at the Mint*, 48–49.
18. Ibid., 23.

19. David A. Schweidel, *Profiting from the Data Economy: Understanding the Roles of Consumers, Innovators and Regulators in a Data-Driven World* (Upper Saddle River, N.J.: Pearson FT Press, 2014), 81.
20. Ibid., 82; Accenture white paper, "City of New York: Using Data Analytics to Achieve Greater Efficiency and Cost Savings," 2013, https://www.accenture.com/t20150624T211456Z__w__/us-en/_acnmedia/Accenture/Conversion-Assets/DotCom/Documents/Global/PDF/Technology_7/Accenture-Data-Analytics-Helps-New-York-City-Boost-Efficiency-Spend-Wisely.pdf.
21. Patrick McGeehan, Russ Buettner, and David W. Chen, "Beneath Cities, a Decaying Tangle of Gas Pipes," *New York Times,* March 23, 2014, A1.
22. 2016 Federal Reserve Payments Study, https://www.federalreserve.gov/paymentsystems/2016-payment-study.htm.
23. Michael Morisy, "How PayPal Boosts Security with Artificial Intelligence," *MIT Technology Review,* January 25, 2016, https://www.technologyreview.com/s/545631/how-paypal-boosts-security-with-artificial-intelligence/.
24. "Brooklyn Nets' Jeremy Lin on New Partnership," television interview on *Squawk Box,* CNBC, February 8, 2017, http://video.cnbc.com/gallery/?video=3000591640.
25. James Ham, "Kings Add New Stat Guru Luke Bornn to Front Office," NBC Sports, April 20, 2017, http://www.csnbayarea.com/kings/kings-add-new-stat-guru-luke-bornn-front-office.
26. Alexander Franks, Andrew Miller, Luke Bornn, and Kirk Goldsberry, "Counterpoints: Advanced Defensive Metrics for NBA Basketball," paper presented at the 9th Annual MIT Sloan Sports Analytics Conference, 2015, http://www.lukebornn.com/papers/franks_ssac_2015.pdf.
27. "Brooklyn Nets' Jeremy Lin on New Partnership," television interview on *Squawk Box,* CNBC, February 8, 2017, http://video.cnbc.com/gallery/?video=3000591640.

CHAPTER 6

1. Gabrielle Glaser, "Unfortunately, Doctors Are Pretty Good at Suicide," *Journal of Medicine*, August 15, 2015, https://www.ncnp.org/journal-of-medicine/1601-unfortunately-doctors-are-pretty-good-at-suicide.html.
2. Mark Bostridge, *Florence Nightingale: The Making of an Icon* (New York: Farrar, Straus & Giroux, 2008), 56–60.
3. Ibid., 35.
4. Ibid., 31–35.
5. Ibid., 68–70.
6. Ibid., 47–50.
7. Ibid., 105.
8. Ibid., 157.

9. Introduction to *The Collected Works of Florence Nightingale*, vol. 14, *The Crimean War*, ed. Lynn McDonald (Waterloo, Ont.: Wilfrid Laurier University Press, 2010), 9.
10. Bostridge, *Florence Nightingale*, 248.
11. Ibid., 219–20.
12. Ibid., 203.
13. Ibid., 220, 225–29.
14. Ibid., 229.
15. Ibid.
16. Letter, August 7, 1855, in *Collected Works of Florence Nightingale*, vol. 14, McDonald, ed., 204.
17. Bostridge, *Florence Nightingale*, 237.
18. Lynn McDonald, "Florence Nightingale and Her Crimean War Statistics: Lessons for Hospital Safety, Public Administration and Nursing," transcript of presentation at Gresham College, October 30, 2014, https://www.gresham.ac.uk/lectures-and-events/florence-nightingale-and-her-crimean-war-statistics-lessons-for-hospital-safety-.
19. Bostridge, *Florence Nightingale*, 248.
20. Ibid., 226.
21. Ibid., 229.
22. *The Times*, February 8, 1855, quoted in E. T. Cook, *The Life of Florence Nightingale*, 2 vols. (London: Macmillan, 1913), 1:236–37, available at https://archive.org/details/lifeofflorenceni01cookuoft.
23. Bostridge, *Florence Nightingale*, 260–62.
24. Ibid., 321.
25. E. H. Sieveking, "Training Institutions for Nurses," *Englishwoman's Magazine* 7 (1852): 294, from Anne Summers, "The Mysterious Demise of Sarah Gamp: The Domiciliary Nurse and Her Detractors," *Victorian Studies* 32, no. 3 (1989): 365.
26. Edwin W. Kopf, "Florence Nightingale as Statistician," *Journal of the American Statistical Association* 15, no. 116 (1916): 390.
27. John Hall, letter to Dr. Andrew Smith, April 6, 1856. The letter was sold at auction in 2007 and is transcribed at https://www.bonhams.com/auctions/15231/lot/26/.
28. Florence Nightingale, "Notes on the Health of the British Army," in *Collected Works of Florence Nightingale*, vol. 14, McDonald, ed., 864.
29. Kopf, "Florence Nightingale as Statistician," 390.
30. Ibid.
31. Bostridge, *Florence Nightingale*, 345.
32. Ibid., 335–39.
33. Florence Nightingale, "Notes on the Health of the British Army," in *Collected Works of Florence Nightingale*, vol. 14, McDonald, ed., 854–55.
34. Kopf, "Florence Nightingale as Statistician," 394.

35. Florence Nightingale, "Notes on Hospitals," in *Collected Works of Florence Nightingale*, vol. 16, *Florence Nightingale and Hospital Reform*, ed. Lynn McDonald (Waterloo, Ont.: Wilfrid Laurier University Press, 2012), 215.
36. Kopf, "Florence Nightingale as Statistician," 397.
37. Jocelyn Keith, "Florence Nightingale: Statistician and Consultant Epidemiologist," *International Nursing Review* 35, no. 5 (1988): 147–50.
38. Jan Beyersmann and Christine Schrade, "Florence Nightingale, William Farr and Competing Risks," *Journal of the Royal Statistical Society, Series A (Statistics in Society)* 180, no. 1 (Jan. 2017): 285–93.
39. Most of the measurements of Joe's kidney function would have been of his serum creatinine, rather than his GFR directly. All these measurements were converted here to GFR for the purpose of visualization. Joseph Futoma et al., "Scalable Joint Modeling of Longitudinal and Point Process Data for Disease Trajectory Prediction and Improving Management of Chronic Kidney Disease," in *Proceedings of the 32nd Conference on Uncertainty in Artificial Intelligence*, ed. Alexander Ihler and Dominik Janzig (Corvallis, Ore.: AUAI Press, 2016), 222–31.
40. No patient data was accessed in creating this plot. This data has been simulated to resemble the data from the actual patient. To find a plot of the original data, please see Futoma et al., "Scalable Joint Modeling," 223.
41. Kevin C. Oeffinger et al., "Breast Cancer Screening for Women at Average Risk: 2015 Guideline Update from the American Cancer Society," *Journal of the American Medical Association* 314, no. 15 (2015): 1599–1614.
42. Letter to William Farr, September 14, 1859, in Lynn McDonald, ed., *Collected Works of Florence Nightingale*, vol. 5, *Florence Nightingale on Society and Politics, Philosophy, Science, Education and Literature* (Waterloo, Ont.: Wilfrid Laurier University Press, 2003), 76.
43. Futoma et al., "Scalable Joint Modeling."
44. J. Balog et al., "Intraoperative Tissue Identification Using Rapid Evaporative Ionization Mass Spectrometry," *Science Translational Medicine* 5, no. 194 (2013): 194ra93; "'Intelligent Knife' Tells Surgeon If Tissue Is Cancerous," Imperial College London press release, July 17, 2013, http://www3.imperial.ac.uk/newsandeventspggrp/imperialcollege/newssummary/news_17-7-2013-17-17-32.
45. "MiniMed 670G System Launches in the United States," Medtronic *Meaningful Information* blog, June 7, 2017, https://www.medtronicdiabetes.com/blog/fda-approves-minimed-670g-system-worlds-first-hybrid-closed-loop-system/.
46. P. J. Schüffler et al., "Semi-automatic Crohn's Disease Severity Estimation on MR Imaging," in *Abdominal Imaging: Computational and Clinical Applications*, ed. H. Yoshida, J. Näppi, and S. Saini (Heidelberg and New York: Springer, Cham, 2014), 128–39.
47. Thomas J. Fuchs and Joachim M. Buhmann, "Computational Pathology:

Challenges and Promises for Tissue Analysis," *Computerized Medical Imaging and Graphics,* vol. 35, nos. 7–8 (2011): 515–30.

48. Varun Gulshan et al., "Development and Validation of a Deep Learning Algorithm for Detection of Diabetic Retinopathy in Retinal Fundus Photographs," *Journal of the American Medical Association* 316, no. 22 (2016): 2402–10. The research partnership with Moorfields Eye Hospital is described in a press release at http://www.moorfields.nhs.uk/news/moorfields-announces-research-partnership.
49. Jim McHugh, "Man, Machine and Medicine: Mass General Researchers Using AI," Nvidia blog, December 7, 2016, https://blogs.nvidia.com/blog/2016/12/07/mass-general-researchers-ai/. See also Lee Bell, "Nvidia to Train 100,000 Developers in 'Deep Learning' AI to Bolster Healthcare Research," *Forbes.com,* May 11, 2017, https://www.forbes.com/sites/leebelltech/2017/05/11/nvidia-to-train-100000-developers-in-deep-learning-ai-to-bolster-health-care-research/.
50. See, e.g., Tom Simonite, "The Recipe for the Perfect Robot Surgeon," *MIT Technology Review,* October 14, 2016, https://www.technologyreview.com/s/602595/the-recipe-for-the-perfect-robot-surgeon/.
51. David Szondy, "IBM's Watson Adapted to Teach Medical Students and Aid Diagnosis," *New Atlas,* October 21, 2013, http://newatlas.com/ibm-supercomputer-watsonpath/29415/.

CHAPTER 7

1. Chris Anderson, "The End of Theory: The Data Deluge Makes the Scientific Method Obsolete," *Wired*, June 23, 2008, https://www.wired.com/2008/06/pb-theory/.
2. C. R. Cardwell et al., "Exposure to Oral Bisphosphonates and Risk of Esophageal Cancer," *JAMA* 304, no. 6 (August 11, 2010): 657–63.
3. J. Green et al., "Oral Bisphosphonates and Risk of Cancer of Oesophagus, Stomach, and Colorectum: Case-Control Analysis Within a UK Primary Care Cohort," *BMJ* 2010;341:c4444.
4. Stephen Jay Gould, "The Streak of Streaks," *New York Review of Books*, August 18, 1988, http://www.nybooks.com/articles/1988/08/18/the-streak-of-streaks/.
5. Brett Green and Jeffrey Zweibel, "The Hot-Hand Fallacy: Cognitive Mistakes or Equilibrium Adjustments? Evidence from Major League Baseball," paper presented at the MIT Sloan Sports Analytics Conference, March 2016, http://www.sloansportsconference.com/wp-content/uploads/2016/02/1422-Baseball.pdf.
6. Thanks to Peter Norvig of Google for this story. We heard him tell this story during a visit to UT-Austin in 2011; it is also told at "On Chomsky and the Two Cultures of Statistical Learning," http://norvig.com/chomsky.html.
7. This is usually attributed to the statistician George Box.

8. See, for example, Edward Beltrami and Jay Mendelsohn, "More Thoughts on DiMaggio's 56-Game Hitting Streak," *Baseball Research Journal* 39, no. 1 (Summer 2010), available at https://sabr.org/research/more-thoughts-dimaggio-s-56-game-hitting-streak.
9. Kimberly Daniels, William D. Mosher, and Jo Jones, "Contraceptive Methods Women Have Ever Used: United States, 1982–2010," *National Health Statistics Reports* no. 62, February 14, 2013, http://www.cdc.gov/nchs/data/nhsr/nhsr062.pdf.
10. See, e.g., Guttmacher Institute Fact Sheet, "Contraceptive Use in the United States," September 2016, https://www.guttmacher.org/fact-sheet/contraceptive-use-united-states.
11. Gregor Aisch and Bill Marsh, "How Likely Is It That Birth Control Could Let You Down?" *New York Times* Sunday Review section, September 13, 2014.
12. James Trussell, "Contraceptive Failure in the United States," *Contraception* 83, no. 5 (May 2011): 397–404.
13. The sole exception is female sterilization, for which long-term data does exist.
14. This standard has been widely adopted in the literature since it was proposed by Trussell and Kost in the 1980s in "Contraceptive Failure in the United States: A Critical Review of the Literature," *Studies in Family Planning* 18, no. 5 (1987): 237–83.
15. Gustave Flaubert, *Correspondance* (Paris: Louis Conard, 1929), 5:111. In the original French the quotation reads: "La rage de vouloir conclure est une des manies les plus funestes et les plus stériles qui appartiennent à l'humanité."
16. Joshua Klein, "When Big Data Goes Bad," *Fortune*, November 5, 2013, http://fortune.com/2013/11/05/when-big-data-goes-bad/.
17. Catherine Talbi, "'Keep Calm and Rape' T-Shirt Maker Shutters After Harsh Backlash," *Huffington Post,* June 25, 2013, https://www.huffingtonpost.com/2013/06/25/keep-calm-and-rape-shirt_n_3492411.html.
18. Silla Brush, Tom Schoenberg, and Suzi Ring, "How a Mystery Trader with an Algorithm May Have Caused the Flash Crash," *Bloomberg News,* April 21, 2015, https://www.bloomberg.com/news/articles/2015-04-22/mystery-trader-armed-with-algorithms-rewrites-flash-crash-story.
19. J. Ginsberg et al., "Detecting Influenza Epidemics Using Search Engine Query Data," *Nature* 457 (February 19, 2009): 1012–14.
20. D. Lazer et al., "The Parable of Google Flu: Traps in Big Data Analysis," *Science* 343 (March 14, 2014): 1203–5.
21. D. R. Olson etal., "Reassessing Google Flu Trends Data for Detection of Seasonal and Pandemic Influenza: A Comparative Epidemiological Study at Three Geographic Scales," *PLOS Computational Biology* 9, no. 10 (2013), https://doi.org/10.1371/journal.pcbi.1003256.
22. Lazer et al., "The Parable of Google Flu."

23. Ibid.
24. Although this story may have an earlier origin, we encountered it first in Hubert L. Dreyfus and Stuart E. Dreyfus, "What Artificial Experts Can and Cannot Do," *AI & Society* 6, no. 1 (1992): 18–26.
25. Julia Angwin and Jeff Larson, "Bias in Criminal Risk Scores Is Mathematically Inevitable, Researchers Say," *ProPublica,* December 30, 2016, https://www.propublica.org/article/bias-in-criminal-risk-scores-is-mathematically-inevitable-researchers-say.
26. Julia Angwin, Jeff Larson, Surya Mattu, and Lauren Kirchner, "Machine Bias," *ProPublica,* May 23, 2016, https://www.propublica.org/article/machine-bias-risk-assessments-in-criminal-sentencing.
27. Leah Sakala, "Breaking Down Mass Incarceration in the 2010 Census," Prison Policy Initiative report, May 28, 2014, https://www.prisonpolicy.org/reports/rates.html.

INDEX

Page numbers in italics refer to figures.